Astronomers' Universe

Other titles in this series

Calibrating the Cosmos: How Cosmology Explains Our Big Bang Universe
Frank Levin

The Future of the Universe (forthcoming)
A.J. Meadows

Stephen Eales

Origins

How the Planets, Stars, Galaxies, and the Universe Began

Stephen Eales
Cardiff University
Department of Physics and Astronomy
Cardiff, CF24 3AA
United Kingdom

British Library Cataloguing in Publication Data
A catalogue record for this book is available from the British Library

Library of Congress Control Number: 2006922569

ISBN-10: 1-84628-401-5
ISBN-13: 978-1-84628-401-4

Printed on acid-free paper.

9 8 7 6 5 4 3 2 1

springer.com

Contents

Preface: An Observer's Manifesto

I have always thought that the title of the most popular astronomy book of all time was a bit of a fraud. Steven Hawking's famous book was mostly about a tiny sliver of time — the first 0.000 000 000 000 000 000 000 000 000 000 000 0001 seconds after the Big Bang. This is an important sliver that is believed to contain the answers to many fundamental questions. Can we construct a theory that will unify the two revolutionary theories, general relativity and quantum mechanics, which were two of the most important scientific discoveries of the twentieth century? Is there even a "theory of everything" that will unify all the forces of nature? However, according to the latest results from the WMAP satellite, the Big Bang occurred — and therefore *time* began — 13.7 billion years ago. Therefore, to write a book that excludes 99.9999 per cent (I will not bother with the remaining 37 digits) of the history of the Universe, including the important part in which planets, stars, galaxies — all the things that are important to us — formed, and then call it *A Brief History of Time* does seem, to say the very least, rather inaccurate.

This is a book about what happened next, especially the origins of the planets, stars, and galaxies. It is a good moment to write such a book because we have probably learned as much about these subjects in the last ten years as we have in all the time before, and much of this recent research has not yet diffused from the scientific journals into the public consciousness. There is also one huge advantage in writing about this later period in the history of the Universe. The earlier period is important because of the big unanswered questions, but it is so long ago that what is written about it is often highly speculative and uncertain. In contrast, we have a surprising amount of very definite and concrete information about most of the rest of the history of the Universe, especially from about 2 seconds after the Big Bang until the present day. For a start, astronomers have the huge advantage over historians, archaeologists, and journalists in that they really can observe history as it is happening. The fact that the speed of light, though very large, is finite means that looking out into space is the equivalent of looking back in time; we can sit on the third planet of our average star and use our telescopes to look at events billions of years in the past. According to the latest results from WMAP, we can observe historical events all the way back to four hundred thousand years after the Big Bang. Before this time, we can not observe events directly because the Universe was ionized, which obsçures our view in the same way that the center of the Sun, a ball of ionized gas, is hidden from our view. However, in the same way that we think we understand the processes in the center of the Sun because nobody has been able to think of any other way of explaining the Sun's exterior properties, we have fairly definite knowledge of events in the Universe at earlier times. In particular, the Universe must have had certain properties about two seconds after the Big Bang to explain the chemical elements we see around us today.

The final part of this book is about the biggest of the origin questions, the origin of the Universe itself. In the book's final chapter, I do travel back to this earlier time. My view of this period, though, is rather different. I am an observational astronomer rather than a theoretical physicist, so I am less interested in (and not an expert in) the theories about this period. I am more interested in gritty facts. What facts do we know about this period and what is speculation? What conclusions can we tease out of the few facts that we do know? Can we build telescopes that will allow us to look even further back toward the Big Bang? This chapter is short on the abstract beauty of theoretical physics, but it does try and give a hard-nosed observer's view of what we know and don't know about the first fraction of a second after the Big Bang.

This final origin question is, of course, different in kind from the other three. It is not even clear whether the question has any meaning. If the Universe is defined as consisting of everything there is, does it really make sense to ask how it began — a question that presupposes the existence of there being something *other* than the Universe. It is impossible to discuss this question without moving far from the comfortable world of an observer — the world of telescopes, stars, and galaxies — into the strange worlds of philosophy and of the meaning of language. It is also a question that has been discussed in many other books. In keeping with the observational slant of this book, I have tried to sift through the speculations of physicists and philosophers for ideas that we might someday be able to test with our telescopes.

I have written this book for a reader without any prior knowledge of science, and I have tried hard not to slip into astronomer's jargon and to explain each technical term as I come to it. One of the challenges of writing any book, popular or otherwise, about research in these fields is the pace of change. This means that by the time this book is in print it will be out-of-date. I have taken out some basic insurance against obsolescence by providing a website to accompany this book, which contains new results obtained since this book was published about all of the origin questions (www.originquestions.com).

One common style of science writing, used in many otherwise excellent books, is to describe the present state of scientific knowledge without much explanation of how scientists arrived at this state. I am not a great fan of this ahistorical style for two reasons. First, it tends to give the impression of the present state of knowledge as something immutable — a finished and polished body of work. In reality, the present state of knowledge is always tentative, and some of the discoveries described in this book will undoubtedly vanish within a few years like the morning dew. Second, this writing style also tends to denude the science of all human personality and leave the impression that science is an activity carried out by disembodied intellects, whereas in reality it is a vigorous human activity. In this book, I have always tried to tell the human story of each discovery. The book is therefore a mixture of a description of our present state of knowledge and an explanation of how this state of knowledge came to be. Occasionally in the book I have also told stories from my own career as an astronomer. This is not because my career has any more significance than the careers of the rest of the several thousand professional astronomers around the world, but because I wanted to give the reader a feeling for what it has been like to *be* an astronomer during this exciting period in our subject's history.

I should immediately add that I do not make any great scholarly claims for the historical parts of this book. My account of the recent research into the origins questions is inevitably biased by my own personal geographical and intellectual trajectory over the last two decades; another scientist would undoubtedly emphasize a slightly different set of discoveries as being the important ones. The book is also biased because I have picked out discoveries that make good stories. The historical parts of this book are probably closer to journalism than real history, but I have at least tried to be a good journalist and get the story of each discovery as straight as possible. Because of the limited amount of written information about many of these discoveries, I have often had to rely on the memories of the participants. I am particularly grateful to David Jewitt for his comments about the discovery of the Edgeworth–Kuiper Belt, Derek Ward-Thompson for his account of the discovery of Class 0 protostars, Phil Mauskopf for his memories of the BOOMERANG project, and Simon Lilly for checking my memories of the *annus mirabilis* in our own research field.

The colleagues who have helped me during my own career as an astronomer are too numerous to mention, but I can at least have the pleasure of thanking the following colleagues for specific help with this book, which has ranged from casual conversations over coffee to reading and making comments on individual chapters: Anthony Aguirre, Elizabeth Auden, Mike Edmunds, Rhodri Evans, Walter Gear, Dave Green, Haley Gomez, Simon Goodwin, Dave Jewitt, Simon Lilly, Malcolm Longair, Phil Mauskopf, Dimitris Stamatellos, Derek Ward-Thompson, and Anthony Whitworth.

I am particularly grateful to Gwyneth Lewis, who was the "idiot reader," as she describes it. Without any scientific background, she read the entire manuscript to check that I was explaining things as clearly as I thought (I often was not). As a professional writer and the official national poet of Wales, Gwyneth also made many invaluable comments about style, language, and the art of writing. Also in the world of writers and publishing, I am grateful to Simon Mitton for his original encouragement to write a book, John Watson for taking a flier on an unknown author, and Harry Blom, Christopher Coughlin, and Louise Farkas at Springer.

I thank my children, Nicholas, Juliet, and Oliver, for a reason that will become clear. Above all, I thank my wife Keirsten. Without her love and support over the last two decades, I would not be an astronomer and would never have written this book. I dedicate it to her.

Stephen Eales
Cardiff, UK

Part I
Planets

So we beat on, boats against the current, borne back ceaselessly into the past.

—F. Scott Fitzgerald

1. Rocks

Every few months I take my children to the National Museum of Wales in the center of Cardiff. We have a strict routine. We start off with the Exhibition of the Evolving Earth. This begins in darkness in a small room lined with screens. There is an explosion of light: the Big Bang. On the screens the Universe rapidly expands, galaxies and stars form out of swirling clouds of gas, and eventually the Earth is formed. We step out of the room into a series of winding galleries displaying the history of the Earth. As we walk through the galleries, always moving forwards in time, we travel through the Silurian and Devonian eras, past fossils of primitive sea life, models of long-extinct giant insects and displays showing how the climate has changed and how what is now land was once under the sea. However, the children never walk. They run forward in time to the exciting bit in the Earth's history: the age of the dinosaurs. The dinosaur gallery has skeletons of both land and sea dinosaurs and the huge fossilized skull of a Tyrannosaurus Rex. Even more exciting than the dinosaur gallery is the Ice Age gallery which comes next; here there is a life-sized model of a woolly mammoth which moves when you break an infrared beam. After the Exhibition of the Evolving Earth, we visit the Natural History Exhibition and pay a call on the shark and the giant sea turtle and, occasionally, if one of the children has been doing a history project at school, we may deign to visit the archaeology section. We avoid the art gallery and the exhibitions of ceramics and postage stamps. We always end the visit with an argument in the cafeteria over the cost of each other's desserts.

Right at the beginning of the Exhibition of the Evolving Earth, on the left-hand side, there is a meteorite that was discovered in Gibeon, Namibia, in 1836. It is about the size of human head and made of iron. It is shaped more like a huge potato than a head, though, and it is covered in bumps about an inch in size. The meteorite looks as if it has been polished because, as it plummeted through the atmosphere, the heat from the friction melted its surface layer. Like most meteorites it is over four billion years old. Every time we visit the museum I touch it, feeling a compulsion to touch something that is so old and has come from space.

Immediately after the meteorite there are three rocks. One is labelled the oldest rock in Wales, the second the oldest rock in Britain and the third the oldest rock in the world. The oldest rock in Wales is 702 million years old. The oldest rock in Britain is from North-West Scotland and is 3300 million years old. The world record holder is from Canada and is 3962 million years old. For me this sequence of three rocks is a vivid reminder that the Earth is not merely the eternal backdrop of our individual human stories but the subject of an incident-packed story of its own.

Another display shows that this story is continuing. This is a dial showing the current distance between Europe and North America to an accuracy of a millionth of a millimeter. The figure on the dial is constantly increasing, showing that Europe and North America are moving away from each other. The reason for this is that the Earth's crust is divided into plates that float on the hot rock underneath. Europe and North America are on two plates that are gradually moving apart. As the plates separate, molten rock flows up from the Earth's interior to fill the gap; at other places rock is being destroyed, as one plate is forced down under another plate until it is melted in the Earth's interior. The motion of the plates is not large, only a few centimeters a year, but over time it adds up – one hundred and fifty million years ago Britain was not at its current chilly northern latitude and was not far from the equator.

After this line of rocks there is for me another talismanic rock. It is in a glass case and is so small, about two inches in size, that I did not notice it for several years. The rock has a light gray color and, if you look closely, there are tiny specks embedded in the rock that glisten under the museum lights.

I wish I could touch this rock. In the late 1960s and early 1970s, the Apollo space missions brought 382 kilograms of rock back from the Moon. This tiny piece of rock, on loan from NASA, is one of the few rocks ever brought from another world.

It is just about possible to see where this rock comes from with the naked eye. The Moon is so much part of the furniture of our lives that its distinctive appearance, the pattern of light and dark that looks like a face, is something we usually hardly notice. After Galileo's discovery with one of the first telescopes that the Moon is not a lump of cheese, a celestial lamp or a goddess, but merely a world like our world, the astronomers of the time decided that the dark areas were probably the Moon's oceans and the light areas its land. With our advanced technology (I can do better than Galileo with a pair of binoculars in my back garden) we can see that they were wrong. The dark areas contain the occasional crater and so cannot be oceans. They are actually flat plains of rock. The light areas are hilly terrain. The light areas are so covered in craters that the edge of one crater is often obliterated by another crater, and there are often craters within craters. As a flat plain seemed the safest place to land, the first Apollo mission to land on the Moon, Apollo 11, landed in the Sea of Tranquillity. The light gray rock in the museum, however, comes from the hills and was brought back by one of the later Apollo space missions, probably Apollo 16.

Apollo. Apollo is to me a numinous word because, looking back across the years, Apollo is probably why I, like many others of my age, became a scientist.

A memory of Apollo. Nineteen sixty eight. This is the year of the Prague Spring, a year in which Russian tanks crushed the liberalizing communist regime in Czechoslovakia, the year in which Richard Nixon became president in the United States, the year in which Robert Kennedy and Martin Luther King were assassinated. It is an ugly year of street protests and political murder, a year in which the optimism of the 1960s turned sour. It is also the year in which a manned spacecraft left Earth orbit for the first time. At the end of the year, Apollo 8 travelled around the far side of the Moon and took the famous pictures, watched in living rooms everywhere that Christmas, of the Earth rising above the horizon, a blue half-circle streaked with white—the first time the world saw the world as a world.

A memory of Apollo. Nineteen sixty nine. I am sitting cross-legged on the floor of the hall of Moor Hall Primary School. The whole school has gathered to watch Neil Armstrong and Buzz

Aldrin step out, for the first time, on the surface of another world. It is not very dramatic. There is a long wait and then two faceless figures descend a ladder. There is a crackly, carefully rehearsed statement* transmitted across a quarter of a million miles of space and out to the waiting TV audience, and then the two figures, bounding in slow motion across the Moon's surface, start doing things with scientific equipment I do not understand. Not much happens, but when the school day ends I run home as fast as possible so that I will not miss anything from the most important event that will take place in my lifetime.

A memory of Apollo. Nineteen seventy one. The world is beginning to get bored. Attempts to enliven the TV coverage by introducing a lunar rover for the astronauts to drive and sport (lunar golf) are not succeeding, and people are beginning to question the expense. I am now in high school and have a friend, Gareth Williams, with whom I have many enjoyable lunchtime debates. One of our topics is the space program. Gareth's argument is that the billions of dollars spent on the Apollo missions could be better spent on Earth, feeding the hungry, housing the homeless, and generally solving the world's problems. I argue that if the money had not been spent on Apollo, it would probably have been spent on guns and missiles rather than anything useful. I could have made an argument based on Apollo's scientific research, but even then I am uneasily aware that the huge cost of Apollo, nineteen billion dollars, is because of the need to take the astronauts safely there and back; much of the scientific program could have been carried out by cheap unmanned spacecraft. Apollo is more a jolly adventure to another world than a sober scientific mission (I actually think the "jolly adventure" argument is also a good one, but I do not think this will appeal to the puritanical Gareth, who I am sure is destined for a life in left-wing politics).

This mixture of public and private memories I can just about justify in a chapter that is supposed to be about the latest research into the origin of the Solar System, because Apollo marked the beginning of the period in which we started to systematically explore our own planetary system. Virtually all we have learned

---

* For those under forty, "That's one small step for man, one giant leap for mankind".

about the planets has been learned since Apollo – within a single human generation. Not only have we been lucky enough to live during a time when humans have set foot on another world for the first time, we have also been lucky enough to live during the great period of planetary exploration.

Admittedly, for someone brought up on science fiction books, the space program after Apollo has been a disappointment because humans have not travelled to the planets. Although science fiction writers from the 1940s and 1950s were too conservative in their predictions for when humans would land on the Moon, they were wildly optimistic about when humans would reach other planets. The year 2000 was a fairly typical prediction for the first landing on Mars, and the millennium has come and gone with the manned space program still mired in low-Earth orbit. Nevertheless, although the exploration of the Solar System has not been the jolly adventure I for one would have liked, it has still been one of the great epochs of discovery in human history. It is also an epoch that is not yet over. As I write, a European spacecraft is mapping Mars in exquisite detail and two American robot geologists are prowling around on the surface of the planet trying to see what it is made of. At the same time, the joint American–European Cassini spacecraft is cruising among the moons of Saturn and has recently launched a probe that has landed on the surface of the largest moon, Titan, the only moon in the Solar System with a substantial atmosphere. I have listed in Table 1.1 some of the important voyages, as I see them, in this great epoch of human discovery.

Although I have not space in this book to describe the exploration of the Solar System in the detail it deserves, I want to describe just one space mission as an example of how much our knowledge of the planets has expanded in a single generation. Until the 1970s the moons of Jupiter remained the points of light discovered by Galileo in 1609. In the early 1970s, scientists at NASA realized that the outer planets – the gas giants Jupiter, Saturn, Uranus and Neptune – were in a configuration that made it possible to send a spacecraft to several planets in one mission; with a careful choice of launch date, the spacecraft would pass by one planet, using the gravitational force of that planet like a slingshot to hurl it on to the next. Before the Pioneer and Voyager space missions to the outer planets, a fair amount was known about Jupiter, which is big enough to study from the Earth, but virtually

Table 1.1 The great epoch of planetary exploration*

1970 (Venera 7, Russian)	Mission to Venus; first successful landing on another planet.
1971 (Mariner 9, USA)	First detailed images of Mars, which reveal Valles Marineris canyon system, huge volcanoes, and channels cut by water.
1974 (Mariner 10, USA)	First (and so far only) mission to Mercury, which produces images of forty five per cent of the planet's surface, revealing a heavily cratered surface like that of the Moon.
1976 (Viking 1 and 2, USA)	Mars mission that carries first experiments to look for life on another planet (unfortunately with ambiguous results).
1973–1989 (Pioneer 10 and 11, Voyager 1 and 2, USA)	First missions to Jupiter and Saturn; first detailed images of the moons of Jupiter, discovery that Jupiter has a ring system.
1986 (Voyager 2, USA)	Voyager 2 visits Uranus, producing the first proper images of the planet (from the Earth, Uranus just looks like a star); the images show that the planet is quite different from Jupiter and Saturn, being blue and rather featureless; ten new moons are discovered.
1986 (Giotto, Europe)	First images of the nucleus of a comet.
1989 (Voyager 2, USA)	Voyager 2 visits Neptune, producing the first proper images of the planet (from the Earth, Neptune just looks like a star); the images reveal a blue planet like Uranus; six new moons and a ring system are discovered.
1990 (Magellan, USA)	The spacecraft uses radar to look through the clouds and map the surface of Venus for the first time.
1995 (Galileo, USA)	Mission to Jupiter; probe launched into Jupiter's atmosphere.
2004 (Cassini–Huygens, Europe, USA)	Mission to Saturn; first landing on the moon of another planet (Titan).
2005 (Hayabusa, Japan)	First landing on the surface of an asteroid
1997–2010 (Mars Global Surveyer, Mars Pathfinder, Mars Exploration Rovers – USA; Mars Express – Europe; plus many more).	Intensive study of Mars as a prelude to a manned mission.
2011–2014 (BepiColombo, Europe/Japan; Messenger, USA)	First missions to Mercury since Mariner 10 forty years before.
2014 (Rosetta, Europe)	Spacecraft will land on a comet for the first time and study the changes in the comet as it travels towards the Sun.

* I have left out many important missions in this brief history. The date given for the mission is the date on which the spacecraft visited the planet rather than the date on which it was launched from the Earth.

nothing about its moons. When Voyager 1 reached Jupiter in 1979 the pictures sent back to the Jet Propulsion Laboratory in Pasadena shocked the waiting scientists and reporters, revealing bizarre worlds beyond the imagination of science fiction writers.

Of the four largest moons of Jupiter, the ones discovered by Galileo, the one that is closest to the planet is Io (Figure 1.1). Io is

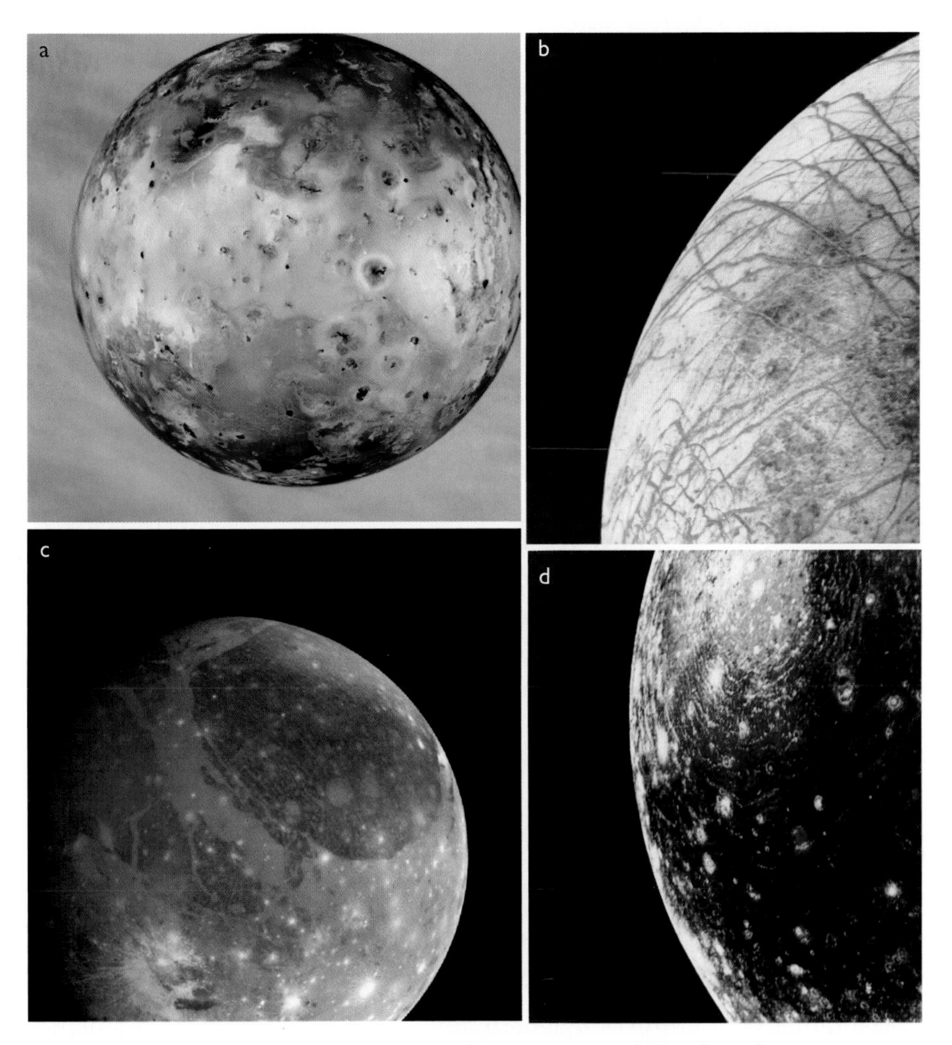

FIGURE 1.1 Montage of black-and-white images of the four largest moons of Jupiter. Io, the innermost moon, is at the top left (in color Io does resemble a pizza); Europa, the second moon, is at the top right; Ganymede, the third moon, is at the bottom left; Callisto, the outermost moon, is at the bottom right. Credit: NSSDC/NASA

about the size of our Moon, but unlike that monochrome world it is a world of vivid color. A journalist, seeing the first image of Io, compared it to a pizza; a scientist said that he did not know what was wrong with the moon but it looked as if it might be cured by a shot of penicillin. The Voyager scientists discovered that Io has more volcanoes per square kilometer than any other world in the Solar System. The volcanoes and the lurid colors are connected. The volcanoes belch out sulphur-rich compounds, which then freeze and fall back as snow onto the moon's surface. Sulphur and chemical compounds containing sulphur have vivid, if not very tasteful, colors, and it is this layer of snow, many meters thick, which is responsible for the moon's bizarre appearance.

The next moon out, Europa, is completely different. The Voyager images showed that it has a smooth, shiny surface covered by a network of fine lines. The NASA scientists realized that the moon must be covered by a thick layer of ice, so thick that the usual topography of a world – the hills, the valleys, the craters – is hidden. The fine lines are cracks in the ice, and the scientists speculated that there might be an ocean under the ice. Twenty years later, the Galileo spacecraft found new evidence for the existence of this ocean*. Because water is one of the basic requirements for life (at least as we know it), Europa's hidden ocean has now risen close to the top of the list of places to look for extraterrestrial life.

The third and fourth moons, Ganymede and Callisto, are also unique worlds but in a more subdued way. The third moon, Ganymede, has strange grooves across its surface and fewer craters than our Moon, which suggests the surface is younger. Callisto, the outermost of the large moons, has a dark surface and is so densely covered in craters that it may have the oldest surface of any object in the Solar System.

The Voyager space mission transformed the moons of Jupiter from points of light into a gallery of worlds. We now understand

---

* The new evidence for an ocean under the ice comes from Galileo's measurements of Europa's magnetic field. The magnetic field of a planet or moon is caused by the motion of electronically conducting material inside the body that is electrically conducting – in the Earth's case, of liquid iron in the Earth's core. Ice does not conduct electricity, water does. Galileo's measurements can be most easily explained if there is an ocean hiding below the ice.

the reason for the differences between them is the gravity of Jupiter. The moons that are closest to Jupiter are so close that the gravitational force exerted by the planet on the near side of the moon is significantly greater than the force on the moon's far side. The difference in Jupiter's gravitational force on the different parts of each moon has the interesting effect that the moon is effectively stretched and squeezed as it orbits around the planet. On Io, the stretching and squeezing heats the center of the moon, in the same way that squeezing and stretching a rubber ball will eventually make it hot; it is this heat that is the cause of the extreme volcanic activity. On Europa, the stretching and squeezing produces the cracks in the surface; on Ganymede, the effect is much weaker, although it is probably responsible for the strange grooves in the surface; and on Callisto, the furthest from Jupiter, there is hardly any effect at all. Although we can now explain these differences as an effect of Jupiter's gravity, without actually visiting the Jovian system, we could never have predicted that this effect would have produced these specific properties – a moon looking like a pizza, for example.

The discoveries of Voyager and the other space missions of the last thirty years are fascinating and awe-inspiring, but they are not fundamental scientific discoveries like Newton's discovery of gravity. In the exploration of the Solar System so far, the geographic and even aesthetic elements have been as important as the purely scientific ones. In a way, the rigorous scientific investigation only starts once the geographical exploration is over. Once we know the properties of the multitude of worlds in our planetary system, we can start to try to answer the question of why the Solar System is like it is? Why, for example, are the four inner planets small balls of rock whereas the next four planets are essentially giant balls of gas? Why do some planets have moons but not others? Why does the Solar System have eight planets (see Chapter 2)? Why does a belt of small objects exist between the orbits of Mars and Jupiter, and why is there another belt of small objects outside the orbit of Neptune? Why is the Earth unique among the inner planets (not only because of the existence of life but also because of things which are not obviously connected to the existence of life, such as the existence of a system of active tectonic plates)? Where do comets come from? How did the Solar System form in the first place?

The most fundamental question is possibly the last one, partly because the answers to some of the other questions would undoubtedly be found in the answer to this one. The question of the origin of the Solar System, and of planetary systems in general, is one of a group of questions often called the "astronomical origin questions." These questions are fundamental scientific questions, but they are also simple ones that have probably occurred to most people. Anyone who has looked at the night sky has probably asked themselves the second of the origin questions: how were the stars formed? It is hard to believe that there is anyone who has never asked themselves the biggest of the origin questions: how did the Universe begin? The remaining origin question is a little less obvious because, at least from the northern hemisphere, one cannot see a galaxy with the naked eye. But whether one can see them or not, galaxies are huge agglomerations of stars (three hundred billion stars in our own) and an obvious question to ask is, how were they formed?

Origin questions are historical questions. A good place to start the discussion of the first origin question, therefore, is deep in the past.

The first person to think seriously about how the Solar System might have formed was the French mathematician, Pierre-Simon Laplace. Laplace was born into a peasant's family just before the French Revolution and ended up his life (demonstrating that a revolution is also a time of opportunity) as the distinguished aristocrat, the Marquis de Laplace*. It is easy to take for granted the properties of the place we live in, the Solar System, but Laplace realized that four of its properties are actually clues to its origin. First, all the planets orbit in the same direction – that is, if you could sit high above the Earth's north pole and look down on the Solar System, you would see all the planets moving around the Sun in the same counterclockwise direction. The second clue is that all the planets (as far as was known at the time of Laplace) rotate on their axes in the same direction. The third is that all the planets orbit around the Sun in the same plane. The final clue,

* Unlike his colleague, the chemist Lavoisier, who lost his head to the guillotine.

again something we take for granted, is that the orbits of the planets are almost circles. Laplace realized these four properties could be explained if the Solar System formed out of a rotating cloud of gas. The cloud would collapse under the influence of gravity, with the collapse occurring along the axis of rotation because of the outwards centrifugal force – the same force that makes it difficult to stay on a merry-go-round. The cloud would therefore collapse into a disk. Laplace suggested that, as the disk of gas cooled, it would break up into rings, rather like the rings of Saturn, with each ring gradually coalescing to form a planet and the material at the center of the disk forming the Sun. This idea explained why the planets are following circular orbits around the Sun and why they are all moving in the same direction. Laplace's remaining clue, the direction of the planets' spin, could be explained by the material at the outer edge of each ring moving slightly more slowly than the material at the inner edge – a prediction of Newton's law of gravitation – which would result in the planet acquiring a spin as it formed out of the ring material. Laplace is known for his highly mathematical and rather dry contributions to a number of sciences, and he was slightly ashamed of his theory, which was not much more complicated than the way I have described it here; he proposed it almost guiltily as a footnote in his five-volume *Mecanique Celeste* "with that uncertainty which attaches to everything which is not the result of observation and calculation."

Scientists have turned Laplace's footnote into a modern theory using two tools. The first tool is the clock provided by the natural process of radioactive decay. This clock has proved invaluable for dating objects in research fields as far apart as astronomy and archaeology. It has, for example, provided the first reliable dates for archaeological sites such as Stonehenge. Our everyday world is made of chemical elements that stay the same. My body is made of carbon, oxygen, hydrogen, phosphorus, potassium, with small amounts of other elements, and these remain carbon, oxygen, hydrogen, phosphorus, potassium, and so on. But there are a few chemical elements that do not remain the same. If I take a lump of pure uranium, leave it for a billion years, and then look at it again, half of the uranium will have turned into lead; if I leave it for another billion years, half of the uranium that is left will have turned into lead; and if I leave it for another billion years, another half of the uranium will have gone – which means that after three billion years, seven eighths of the original uranium will have turned into lead. This transmutation of elements occurs because the uranium atoms are unstable: every now and then (exactly when is a matter of chance) the nucleus of a uranium atom emits a particle and turns into the nucleus of a lead atom. Although the decay of an individual nucleus cannot be predicted, it is possible to predict the behavior of a large enough number of nuclei – that on average a certain percentage of the uranium nuclei will turn into lead nuclei each second. Turning this all around, if I am given a lump of uranium mixed with lead, by knowing how fast uranium transmutes into lead, I can estimate how old the lump is*.

The dates of the rock in the museum come from this technique. The ages of the rock brought back by the different Apollo missions are generally greater than the ages of rocks on the Earth. The dark rock from the lunar "oceans" is between 1700 and 3700 million years old; the lighter rock from the hills is about 4000 million years old. Thus the formation of the Solar System must have occurred at least 4000 million years ago. One problem though with looking at rock from large objects like the Moon and the Earth is that geological processes can melt the rock and reset the radioactivity clock. Better objects for dating the origin of the Solar System are meteorites. Some meteorites are the debris left over after all the ice in a comet has melted and others are probably fragments of rock produced from the collision of asteroids, objects that orbit the Sun

---

* I have simplified things slightly. As I have described it here, this technique will only work if one knows that the lump was originally completely made of uranium, the "parent element". In reality, the lump might well have contained some lead, the "daughter element". The radioactivity clock can still be used, however, as long as there are two different kinds of lead, one of which is formed by radioactive decay from uranium and one of which is not. I have not space to describe the full technique in detail but, briefly, by looking at the ratios of parent to daughter and sister to step-sister in different minerals within a lump of rock, it is possible to estimate both the age of the rock and its original composition.

between the orbits of Mars and Jupiter. Both comets and asteroids are small enough that the clock should not have been reset. Many meteorites have virtually the same age, 4600 million years, which means that the Solar System must be at least this old.

The second tool that scientists have used to expand Laplace's footnote is the computer. The reason that Laplace, who was one of the greatest mathematicians of all time, did not do any calculations himself is that the processes occurring in the disk, often called the solar nebula, were horribly complicated, far too complicated to calculate in the traditional way with pen and paper. Instead, modern scientists use computers to simulate the processes. The problem with computer simulations is that the limitations of computer power mean that the scientist usually has to make some choices about which are the important processes and which ones can be safely left out. Different scientists make different choices and so different simulations produce slightly different results, but they do all produce something that looks like a real planetary system (Figure 1.2).

FIGURE 1.2 Simulation of the formation of a planetary system by Phil Armitage and Ken Rice at the University of Colorado. The bright spots are planets. Credit: Phil Armitage

The standard model of the formation of the Solar System goes like this. Four thousand six hundred million years ago, just before the Solar System was born, there was a rotating cloud of cold gas with a total mass of about one thousand billion billion billion kilograms. Because of the gravitational force exerted by all this gas, the cloud started to collapse under its own weight. When anything falls through a gravitational field, gravitational energy is transformed into other kinds of energy. When a diver dives off a high board, for example, gravitational energy is first transformed into the energy of motion, kinetic energy, and then, when the diver hits the water, into heat. For exactly the same reason, as the cloud collapsed, the gas became hotter. As Laplace first realized, the endpoint of this collapse was a rotating disk of hot gas, the pressure in the hot gas stopping any further collapse.

The Sun formed at the center of the disk, but this is a story for another chapter (Chapter 5). In the rest of the disk, the hot gas began to cool. As the temperature fell, bits of gas began to freeze; first material with a high melting point, metals such as iron, titanium and magnesium; then material with a lower melting point such as water, ammonia, and even carbon dioxide. The temperature would have been higher at the center of the disk close to the Sun, and so the solid material there was mostly made of material with a high melting point.

The story so far is fairly clear. The disagreements are about what happened next. Over the next twenty million years, the tiny solid particles in the disk combined to form planets, but the disagreements are about how they did this. There is some consensus among astronomers that the particles probably initially stuck together in the same way that ice particles coalesce to form snow flakes. The first solid objects to coalesce in this way can probably still be seen. A certain type of meteorite, the carbonaceous chondrite, contains irregular lumps of white material about a centimeter in size. Because these lumps are composed of minerals rich in calcium, titanium and aluminium, all substances which freeze at over 1000 degrees Centigrade, these lumps were probably among the first objects that coalesced out of the solar nebula (Thus when I touch one of these meteorites, I am not only touching something that has come from space, I am touching something from a time before the planets even existed).

There is also some consensus that at some point in the story, the Solar System was filled with objects about 100 kilometers in size – *planetesimals* or little planets. Once planetesimals are present in a simulation, the production of planets is inevitable, because gravity gradually draws the planetesimals together. The planetesimals that are left in a simulation after the planets have been formed provide a natural explanation of the small bodies in the Solar System, such as the asteroids and comets. The radioactivity clocks in all objects would have been reset at this time because of the heat produced by the collisions of planetesimals, and so the standard model nicely explains the common age of many meteorites.

The details of how the small lumps coalesced to form planetesimals are still unclear, but the biggest disagreement between astronomers is about how the giant planets were formed. As one travels out through the Solar System from the Sun, the first four planets – Mercury, Venus, Earth and Mars – are essentially small balls of rock; atmospheres and oceans are important to us but most of the inner planets, including the Earth, is solid rock. The next four planets – Jupiter, Saturn, Uranus, Neptune – are much bigger than the first four (Jupiter would contain 1400 Earths) and are mostly balls of gas. Most astronomers accept that because the inner planets are essentially balls of rock, they must have formed from the coalescence of rocky planetesimals. The disagreement is about how the giant planets formed.

For many years the most popular theory for the formation of the gas giants has been the core-accretion theory. According to this theory, the formation of the gas giants would initially have occurred in the same way as the formation of the inner rocky planets. Gravity would gradually have drawn planetesimals together, leading to the formation of larger and larger objects. However, according to this theory, once the protoplanet, or "core," reached a mass about 15 times the mass of the Earth, its gravitational influence would have become so large that it would have quickly swept up most of the gas in that part of the solar nebula. The mass gained by the planet during this accretion phase would have been at least ten times greater than the mass of the core. The competing theory, which has recently come back into fashion, is the "gravitational instability" theory. According to this theory, the two types of planets were formed in different ways. The inner planets were formed by the coalescence of planetesimals; the gas giants were formed by the sudden gravitational collapse of large parts of the solar nebula – in much the same way that the Sun was formed as the result of the gravitational collapse of a larger cloud of gas. A circumstantial argument in favor of this theory is that the gas giants with their extensive systems of moons do look rather like mini-Solar Systems.

There is no consensus about which of these theories is correct[1]. The big difficulty in deciding between the two is the complexity of the physical and chemical processes involved in forming planets, which means that although computer simulations do produce results that look like real planetary systems, there is no "killer simulation" with the sophistication and complexity necessary to convince astronomers that one of these two theories must be right.

Despite disagreements about some details, most astronomers believe, partly because there is no plausible alternative, that the standard model is correct. A circumstantial piece of evidence is the widespread existence of planetary systems (Chapter 3), because the standard model implies that a planetary system should be formed just about whenever a star is formed. There is also a nice piece of evidence for the standard model from the Voyager missions to the outer planets that I described above.

Apart from producing spectacular pictures, such as the ones in Figure 1.1, the Voyager space missions also made basic measurements of the masses and densities of the moons. These revealed that the apartheid between the planets in the Solar System also applies to their moons. The densities of the moons of the outer planets are generally much less than the density of our own, the only large moon in the inner Solar System. The densities are so low that it seems certain that a significant fraction of the outer moons is ice rather than rock. The standard model explains this rather well. In the inner parts of the disk, the heat from the Sun would have meant that the temperature was too high for water to freeze, and so the inner planets and moons were formed out of particles of rocky material, which freezes at a much higher temperature. In the colder outer part of the disk, ice particles would have been able to form as well, and so one would expect moons in the outer Solar System to have a higher ice fraction, as they appear to do.

Thus the standard model is definitely one part of the story of the Solar System, but it cannot be the whole story. There are a few anomalous facts about the Solar System that it cannot explain.

One of these anomalies is that Laplace was wrong about one thing: all the planets do not spin in the same direction. Two planets actually spin in different directions. The first of these, Venus, is almost the twin of the Earth, having virtually the same mass and diameter. This similarity and the thick clouds hiding its surface made it for many years a favorite location for science fiction writers; it was possible to imagine that there might be Earth-like life hidden under the clouds – dinosaurs crashing around in primeval swamps was one idea – which it wasn't for some of the more obviously hostile planets. However, in the 1960s, when the Russian Venera spacecraft descended through the clouds, it was soon discovered that Earth life transported to Venus would be immediately killed in at least four different ways: asphyxiated by the lack of oxygen; broiled by the high temperature; crushed by the high pressure (seven hundred times that of the atmosphere on Earth); and finally dissolved by the soft rain of sulphuric acid which drizzles down from the Venusian sky. The clouds also made it impossible for a long time to determine in which direction, and how quickly, Venus is spinning. When Venus's rotation was finally measured by radar, it was discovered that Venus rotates much more slowly than the Earth and in the opposite direction; on Venus the Sun rises in the west and it will be 243 Earth days before night falls. Venus, at least, has its axis of rotation roughly parallel to that of the Earth and most of the other planets. The second anomalous planet, Uranus, is not only rotating in the opposite direction but is also lying on its side; its axis is almost at right angles to the axes of the other planets.

Another anomalous fact is the piece of furniture in the sky. Moons are quite common in the Solar System – Jupiter has 63, at last count, ranging in size from the moons discovered by Galileo down to unnamed moons only one kilometer across – but our moon, the Moon, is rather unusual. The other large moons in the Solar System are still much smaller than their planets, but the Moon is an anomaly because it is a solitary moon and because it is so large relative to the Earth.

The Moon may have some connection to life on Earth. In the 1990s, a group of French astronomers suggested that the existence of the Moon was responsible for the relative stability of the Earth's climate[2]. They argued that the gravitational effect of the gas giants caused the axes of the inner planets to move about chaotically (perhaps explaining the anomalous rotation of Venus) but that the axis of the Earth was stabilized by the presence of the Moon. If true, this would have obvious consequences for life (suppose the Moon were not there and imagine the effect on a tribe of primitive humans in Africa if it suddenly found itself moved up to close to the North Pole). It has also been suggested that the oceans' tides, which are caused by the gravitational field of the Moon, may have been responsible for the first colonization of the land by life from the oceans. Whether or not these ideas are correct, in the rest of this chapter I will show that the *process* responsible for the existence of the Moon is almost certainly responsible for our existence as a species. A good place to start the rest of the story is once again with some rocks.

Every now and then I visit the National Museum during the week. The museum is just around the corner from my office and I sometimes spend a leisurely lunch hour wandering around it. The Exhibition of the Evolving Earth is my favorite place in the museum, and without the children it is actually possible to read the labels on the exhibits. As I walk through the twisting dark galleries, I move forward through the history of the Earth, beckoned onwards by the distant sound, sometime in the Cretaceous era, of a bellowing Tyrannosaurus Rex. The narrow galleries themselves have been designed to resemble a serpentine tunnel carved through rock and are littered with rocks and fossils from different chapters in the story of the Earth. Most of the rocks are lumps of sedimentary rock, rocks such as sandstone and limestone that have been formed over millions of years by infinitesimally slow geological processes. The red sandstone of the Devonian era, for example, the period of the first fish, is simply compacted sand, grains of sand that have been compressed and fused together by the weight of the sand on top. These processes continue today, and a cloud of sand kicked up on the beach by the foot of a child may one day be frozen under the Earth as a lump of sandstone.

William Smith is a name that I suspect only one out of a thousand people would recognize. Yet he virtually founded the science of geology and was the first to recognize that the story of the Earth is contained in its rocks.

Smith earned his living as a surveyor (he never made any money as a geologist) and he spent most of his working life in the late eighteenth century overseeing the construction of the English canals. This was before the time of dynamite and mechanical excavators, and the canals were dug, literally, with pickaxe and spade by thousands of "navvies." The navvies would often find fossils when they were digging through sedimentary rock and they would bring these to Smith to look at. He noticed that the types of fossils changed gradually from the lower older layers of rock to the higher newer layers. Sitting in his tent, inspecting the fossils brought by the navvies, Smith must have been frustrated not to understand *why* the fossils changed from layer to layer. We do, of course, because of the work of Darwin. Time deposits the sedimentary rock while it also, through *natural selection*, causes some species to become extinct and new species to arise – and so the species that become entombed in the higher sedimentary layers are different from those in the layers below. But even though Smith did not understand the reason for these changes, he was the first to realize that the story of the Earth is written in sedimentary rocks.

These changes are usually quite gradual; the types of fossil found in one sedimentary layer are usually only slightly different from those in the layer below. But there are some places in the "fossil record" where the change is much more sudden. One of the most spectacular of these jumps in the record is at the boundary between the Cretaceous and Tertiary periods sixty five million years ago*. The fossils that are found in the sedimentary rock deposited at the end of the Cretaceous period are very different

---

* *Cretaceous* and *Tertiary* are now fairly meaningless terms, the coinage of Victorian geologists. *Cretaceous*, for example, comes from the Latin for chalk, *creta*, because chalks were deposited in shallow seas over a large area during this period. The symbiosis of geology and paleontology is shown by the reason for the positioning of the boundary. The rock on either side of the boundary is not actually any different, the boundary being chosen because of the jump in the fossil record at this time.

from those found in the sediments deposited only slightly later at the beginning of the Tertiary period. In the rock layers at the beginning of the Tertiary, about half the species that were present at the end of the Cretaceous period have suddenly vanished. The coiled-shell ammonites, which flourished in the oceans of the Earth for hundreds of millions of years and which are the staple of any museum fossil collection, are not present at all in the rock layers at the beginning of the Tertiary. Many other species of sea creature also vanish from the fossil record. A wide variety of trees and plants also disappear; most types of bird vanish; and, most spectacularly of all, the dinosaurs, lords of the Earth for one hundred and sixty million years, disappear from history.

There have been five major extinctions in the Earth's history. The extinction at the Cretaceous–Tertiary boundary (confusingly usually called the KT boundary after the German term for the boundary) was not the largest extinction – that occurred 250 million years ago at the close of the Permian period – merely the most famous one because it marked the end of Tyrannosaurus Rex and the other dinosaurs. Words like *suddenly* and *vanished* suggest that the KT extinction was an instantaneous event, but for many years the orthodox geological view was that the KT extinction was only sudden when looked at in the context of the hundreds of millions of years of life on Earth.

Consider the case of the disappearing T-Rex. The forensic evidence in the case is very poor. There are only a few body parts spread over tens of millions of years of history (the National Museum is lucky to have a skull). There are T-Rex fossils in the Cretaceous sediments and none in the Tertiary sediments, but the evidence is insufficient to tell whether the crime occurred instantaneously or over the million-year period during which the last sediments in the Cretaceous period and the first sediments of the Tertiary period were laid down. Until the late 1970s, the view of most geologists was the conventional "uniformitarian" one that all geological change occurs gradually. The KT extinction might look sudden in the fossil record but it had probably occurred over a period of several million years, as the result of either climate change or a change in the sea level.

At the end of that decade a young American geologist, Walter Alvarez, began to think the conventional view might be wrong.

Alvarez was studying a particularly good history book: the Scaglia rossa limestone outside the small medieval town of Gubbio in Italy. Gubbio is in the Appenine Mountains and the beautiful red limestone (*rossa* refers to the red color) is found on the sides of the steep valleys outside the town. The limestone is now in the mountains, but millions of years ago it was deep under the ocean; it was laid down on the ocean floor by the slow precipitation of grains of the mineral calcite out of sea water and it is only subsequent upheavals in the Earth's crust that have brought it into the mountains. The limestone beds are as much as 400 meters thick and so form a continuous record of the history of life on Earth covering hundreds of millions of years. The limestone started out under the sea and so the fossils it contains are fossils of sea creatures. A particularly abundant fossil found in the limestone are the *foraminifera – forams* for short. These are tiny floating organisms, which when they die sink to the bottom of the ocean and become incorporated in the sedimentary rock. They do not have the T-Rex problem of only a handful of skeletons spread over millions of years of sediments; even a small piece of limestone can contain several hundred fossil forams.

In the book that he wrote about his discovery, Alvarez describes how while he was carrying out an important but routine scientific investigation – a study of the magnetic properties of the Scaglia rossa limestone – his imagination became gripped by the question of why the dinosaurs had died out. When he looked at the KT boundary in the limestone, he discovered that some large species of forams are found right up to the very edge of the Cretaceous layers but are not present at all in the Tertiary layers (Figure 1.3). He decided that this really did look like a sudden extinction; the patchy history provided by dinosaur bones might be consistent with a gradual extinction over several million years but the abrupt change in the foram fossils looked as if something had happened suddenly. He also noticed that between the last layer of Cretaceous limestone and the first layer of Tertiary limestone there was a layer of clay, about one centimeter thick, which contained no fossils at all. He wondered whether this layer had anything to do with the KT extinction. He wondered whether there was any way of estimating how long it had taken for this clay layer to be deposited.

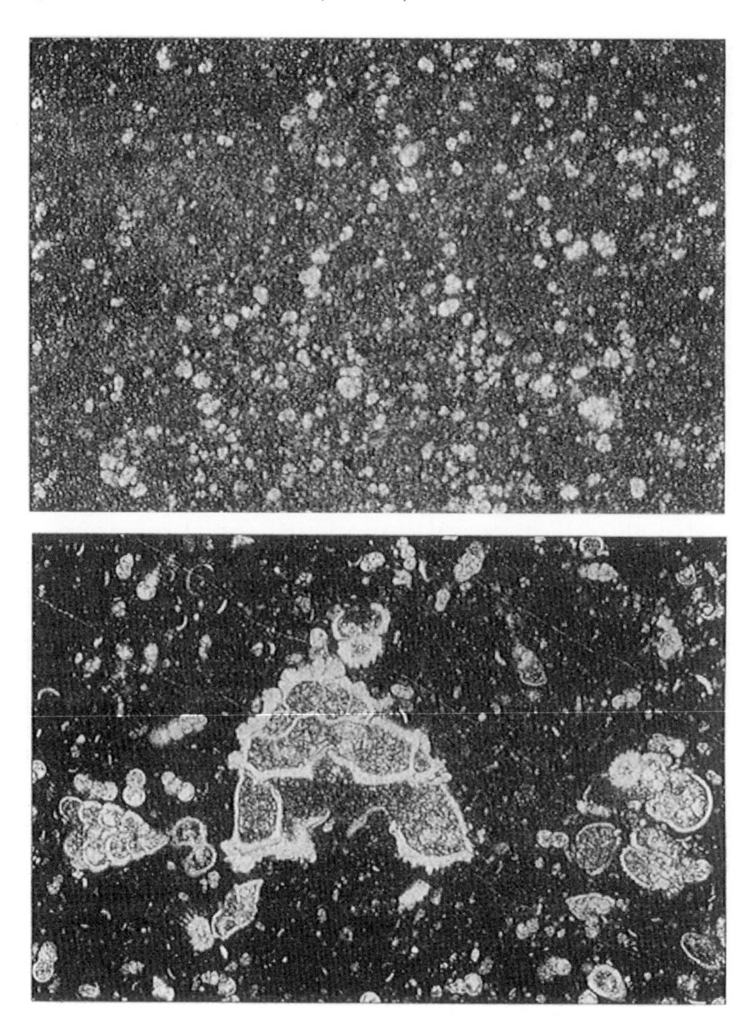

FIGURE 1.3 Photographs, taken through a microscope, of the rock right at the end of the Cretaceous period (bottom) and at the beginning of the Tertiary period (top). The large objects visible in the lower picture are Foraminifera. By the beginning of the Tertiary period they have vanished. Credit: Walter Alvarez

At this point almost every other geologist on the planet would have become hopelessly stuck. However, Walter Alvarez's father was the physicist, Luis Alvarez. The older Alvarez was not only a man of deep insight in his own subject – he had won the Nobel Prize in 1968 – but a man with interests far outside physics. He and his friend, the Egyptian archaeologist Ahmed Fakhri, had once x-rayed the pyramid of Kephren at Giza using cosmic-ray muons, subatomic particles that are constantly bombarding the Earth (They hoped to discover hidden chambers full of treasure, but they disappointingly found the pyramid is solid rock from top to bottom). Walter Alvarez told his father about the KT extinction and the clay layer, knowing that this was exactly the kind of big problem that would catch his imagination. Luis Alvarez started to think about a way to estimate how long it had taken the clay layer to form.

After a number of false starts, he came up with the idea of meteoritic dust. Apart from large meteorites like the one in the museum, the Earth is also constantly being hit by tiny grains of space rock. This constant drizzle of dust from space, which slowly accumulates on the Earth's surface, provides a kind of clock. Alvarez realized that if he could discover a method of measuring the amount of meteoritic dust in the clay layer, he could find out how long it had taken the clay to be deposited – slowly and the clay would contain a large amount of dust; quickly and it would contain hardly any at all. Fortunately, he already knew a method. There are some chemical elements that are rare in Earth rock but are relatively common in asteroids and meteorites. One of these is the element iridium. By measuring the amount of iridium in the clay layer, it would be possible to estimate how much meteoritic dust it contained. Alvarez calculated that if the clay layer had been deposited over several thousand years, about one atom in every ten billion would be an iridium atom; if the clay layer had been deposited suddenly there would be no iridium at all. Of course, detecting one atom out of ten billion is a huge challenge, roughly equivalent to finding one person among the billions of people living on the Earth. But Alvarez knew someone who had developed a method of measuring such minuscule amounts. Fortunately, this man, Frank Asaro, worked at the same university, Berkeley, as both the Alvarezs. Walter Alvarez gave Frank Asaro some samples from the clay layer and then both father and son waited – many months since Asaro's method was extremely complicated and time-consuming.

After almost ten months, Walter Alvarez received a phone call from his father: Frank Asaro had completed his analysis and something was wrong. After months of analysis, after careful checking and rechecking, he had discovered that the iridium content of the clay layer was much higher than expected. Out of every ten billion atoms, ninety were iridium atoms, ninety times more than expected even if the clay layer had been laid down over several thousand years. What could explain such a peculiar result?

Their first idea was that the KT extinction had been caused by radiation from a nearby exploding star, a supernova. But measurements of other elements in the clay, which should have been present if the supernova idea was correct, quickly ruled this out. They then came up with the following idea.

Suppose that sixty five million years ago a small comet or asteroid, about 10 kilometers in size, hit the Earth. Although 10 kilometers does not sound particularly large, the heat produced by such a collision would have been the equivalent of one hundred million hydrogen bombs, enough to kill every living creature within hundreds of kilometers of the point of impact. The collision by itself would not have been enough to wipe out whole species, however – this would have been done by the horsemen of the apocalypse following the initial impact.

The impact would have thrown fragments of rock and dust into the Earth's atmosphere. These would have blocked out the light of the Sun, and for many months the surface of the Earth would have been cold and dark. The foundation of life on Earth is photosynthesis, the process through which plants tap into the energy of the Sun. In the dark months after the impact, photosynthesis would have shut down completely. Vulnerable species higher up the food chain would have starved. After several months, the dust would have settled and light would have returned. But then the cold would have been replaced by heat, the result of greenhouse gases released from rock by the heat of the impact. Many of the species that had survived the long night would have died then. The final horseman would have been acid rain, another consequence of the heat of the impact. Cold, starvation, heat, acid falling from the sky – all of these may have had a role in the Great Extinction.

But did this actually happen? It was the only idea that the Berkeley group could think of that could explain both the KT extinction and the iridium-rich clay layer. As the dust settled back on to the Earth's surface, it would have naturally have formed a thin layer at the boundary between the Cretaceous and Tertiary sediments. This layer would have been rich in iridium because some of the dust thrown into the atmosphere would have come from the asteroid itself*. They proposed this idea in an article in the international magazine *Science* in 1980. However, the idea was not immediately universally accepted because the Berkeley group could not answer one simple question. Where was the crater made by the impact?

An asteroid ten kilometers in size should make a crater about 40 kilometers deep and between 150 and 200 kilometers in diameter – not an easy one to miss. But there is no crater this large on the Earth's surface. This did not disprove the theory because there are ways such a crater might have been hidden. The impact might have occurred in the ocean, and it was still possible, in the early 1980s, that there was a crater this large in one of the unexplored parts of the ocean floor. It was also possible that the Earth itself had concealed the crater. The Earth's crust is divided into plates, which are created by hot rock welling up in the middle of oceans and destroyed when one plate is forced down into the Earth's mantle by a second plate; it was possible that the impact had occurred on a plate which had subsequently been destroyed. It was also possible that, during the sixty five million years since the impact, the crater had gradually been concealed by the deposition of sedimentary rock. Nevertheless, although it did not disprove the theory, the absence of a crater was a distinct embarrassment.

For almost a decade, geologists looked all over the Earth for a large crater formed sixty five million years ago. By the end of the

---

* In their article in *Science* the Berkeley group estimated the size of the asteroid using four different methods. All gave similar answers. One method, to give an example, was to estimate how much iridium there is in the clay layer. The clay layer can be found anywhere on the Earth's surface where rock of the correct age is accessible. It is reasonably easy to make an estimate, albeit imprecise, of the total amount of iridium in the clay. With the assumptions that all this iridium came from the asteroid and that the percentage of iridium in the asteroid was the same as that in meteorites, the Berkeley group was able to estimate the mass of the asteroid and thus its approximate diameter.

1980s they were beginning to suspect that the impact had not occurred in the ocean but on the land. At a number of sites in North America the KT clay layer contained grains of quartz with an unusual structure that looked as if it had been caused by a shock wave. But despite the evidence of the shocked quartz and the iridium, many scientists were still not convinced by the impact hypothesis and during this decade any academic conference on the KT extinction was sure to be the scene of a vigorous debate between the believers and the sceptics. The final answer, though, lay not in the polite world of academic science (even the most vigorous scientific debate is disappointingly tame) but in the much tougher world of oil exploration.

A standard way of looking for oil is to look for gravitational anomalies. Although the gravitational field of the Earth is pretty much the same everywhere on the Earth's surface, it does vary slightly from point to point: higher where there is a dense bit of rock under the surface; lower where the rock under the surface is not very dense. In 1950 geologists working for the Mexican state oil company, PEMEX, became excited by a circular pattern of gravitational anomalies centered on the town of Puerto Chicxulub in northern Mexico. The pattern was about 300 kilometers across and the geologists suspected that it might indicate a gigantic reservoir of oil. They set up a drill to look for the oil but, after penetrating through a kilometer of sedimentary rock, the drill began bringing up hard dense crystalline rock – not the kind that contains oil. For an oil geologist, once a rock structure has been shown not to contain oil it is no longer of much interest. The PEMEX geologists casually concluded that the circular pattern probably marked a huge volcano that had been buried by layers of sediments. Thirty years later, however, two more PEMEX geologists, Antonio Camargo and Glen Penfield, carried out a new survey of the region. This time they concluded that the geological results were actually better explained if the circular pattern marked a buried impact crater. To an oil company, of course, craters are no more interesting than volcanoes; Camargo and Penfield gave one brief talk about their work in 1981 and that was that.

This was only a year after the Alvarezs' article. Camargo and Penfield had not read it, and nobody interested in the KT extinction heard their talk. It was not until nearly ten years later that a

Canadian geologist, Alan Hildebrand, found out about the circular pattern of gravitational anomalies, put two and two together, and realized the PEMEX geologists might have discovered the KT crater. The crater was definitely big enough, but was it made at the same time as the KT extinction?

The answer lay in the rock brought to the surface forty years earlier by the PEMEX drills. Unfortunately, it was initially thought that the old drill cores had been stored in a warehouse that had been destroyed by fire. Glen Penfield thought that there might be some drill cores lying around the old drill sites, but oil companies are rather untidy (especially when oil is not discovered) and nobody could remember precisely where the drill rig had been. Penfield spent some time digging in vain through piles of pig manure where villagers thought the drill rig had been erected forty years earlier. Finally, however, the old cores were found. Radioactive dating of the rock showed that the crater was formed sixty five million years ago. It now seems almost certain that the crater was made by the asteroid responsible for the KT extinction.

The reason I have told this story at some length is that the KT extinction is the closest example of a crucial historical process. The standard model for the formation of the Solar System that I described above is, in geological terms, a uniformitarian theory, because it implies that the Solar System formed by gradual processes: the slow cooling of a disk; the freezing of solid particles out of the disk; and then the gradual coalescence, over twenty million years, of the solid particles into planets. However, it has now become obvious that many of the present properties of the Solar System are not the consequence of gradual processes occurring in the solar disk four-and-a-half billion years ago, but are instead the consequence of sudden events, of impacts coming out of a clear sky – of chance.

Of course, one only has to look up into the sky to see the effect of impacts. The surface of the Moon is scarred with tens-of-thousands of craters, most of which were formed during the first half billion years after the formation of the planets when there were many more rocks flying around the Solar System than there are today. A small telescope or a pair of binoculars is necessary to see the craters, but even with the naked eye it is possible to see the Moon's distinctive pattern of light and dark, the face of the

Man in the Moon. It is now clear that this face is the result of chance.

One of the few tangible results of the Apollo program was the 382 kilograms of Moon rock it brought back. A truck-load of Moon rock does not actually sound very good value for nineteen billion dollars, but there are many kinds of geological and chemical analyses possible in large well-equipped laboratories back on the Earth that are not possible in space vehicles. The careful analysis of all this rock in the decades since the last Apollo mission has gradually revealed the history of the Moon. As I described earlier, geologists have used the radioactivity clock to show that the rock from the dark areas of the Moon is generally younger than the rock from the light areas. The rock from the dark areas is also denser and has a different mineralogical composition - it resembles the rock found in lava flows on the Earth. These clues have led to the following story for the birth of the Man in the Moon.

When it was first formed the Moon was probably so hot that it was a ball of liquid rock. As the rock cooled, a crust of solid rock formed around the molten core. Rock with a low density floated to the surface of the molten ball and so the crust was made out of low-density rock (the light areas of the surface we see today are probably parts of this first crust). Among the many rocks that hit the Moon's surface during the first billion years or so, there were a few particularly large rocks. These hit the surface with such force that the impacts excavated huge basins, basins that were so deep that the crust beneath them was very thin. The denser liquid rock beneath forced its way out through the thin crust and flowed over the surface to form a smooth plain of rock – a lunar "ocean."

This story nicely explains the differences between the light and dark areas on the Moon. Because the basins were excavated by a handful of big rocks, it also implies that the face of the Man in the Moon was a matter of chance. If the roulette wheel of impacts had spun a different way, we would now see a different "face" or even none at all (the far side of the Moon has very few of these dark areas).

Once the importance of chance is acknowledged, one can see its effects everywhere in the Solar System. If the standard model were the entire story, all the planets should be rotating in the same direction. However, we can explain why Uranus is rotating on its side if just after it was formed, it was hit one final time by a very large object. This might also be the explanation of the anomalous rotation of Venus, although for Venus there are other possible explanations of its retrograde rotation[3]. The systems of moons around the planets also seem to be partly a matter of chance. Many of the moons of the giant planets were almost certainly formed from the coalescence of material in disks around the newly formed planets, similar to the process by which the planets themselves were formed. However, some of the smaller moons and also Triton, the largest moon of Neptune, are orbiting their planets in the opposite direction to the other moons. It seems almost certain that these moons were formed elsewhere, and were subsequently captured by the gravity of the planet as a result of chance encounters – if the roulette wheel had spun a different way, these encounters might not have occurred. Some of the moons appear to have been casualties from these early spins of the roulette wheel. The appearance of Miranda, one of the moons of Uranus, is so difficult to explain with standard geological ideas that scientists have suggested that it was once completely broken into pieces by an impact and then roughly plastered together by the force of gravity[4] (Figure 1.4). On this roulette wheel, however, one moon appears to have been a winner.

Before Apollo, there were three theories for the origin of the Moon: the capture theory; the fission theory; and the double-planet theory. In the capture theory, the Moon was formed elsewhere in the Solar System and was then snared by the Earth's gravitational field. The fission theory, which was first suggested by George Darwin, one of Charles Darwin's ten children, was that the Earth had originally been spinning so fast that it bulged at the equator; a blob of material then broke off to become the Moon. The double-planet hypothesis was that the Moon and Earth formed at the same time out of a single cloud of dust and gas. There were problems with all these theories. The capture theory would only work if the Moon had been on an improbably precise orbit – anything slightly different and it would either have collided with the Earth or have been thrown far away from it. The problem with the fission theory was that for material to have broken off, the Earth would have had to be rotating at least once every two and a half hours, and scientists did not think the Earth had ever been

FIGURE 1.4 Image taken by Voyager 2 of Miranda. The surface of the moon is composed of several distinct geological regions. One possible explanation of the surface is that the moon was at one time broken into pieces, and then roughly reassembled by gravity. Credit: NASA/JPL-Caltech

spinning so fast. There were two problems with the double-planet hypothesis. It could not explain why Venus, otherwise very similar to the Earth, does not have a moon and it also could not explain why the Earth has a dense core but the Moon does not.

The rock brought back by Apollo generated a fourth theory. The little piece of Moon rock in the National Museum is different in one important respect from the big lumps of Earth rock surrounding it. Volatile substances, substances with low boiling points such as water but also metals like potassium and sodium, are much scarcer in Moon rock than in Earth rock. On the other hand, refractory substances, substances with high boiling points such as silicon and aluminium, are more common on the Moon than on the Earth. This difference in composition is clinching evidence that two of the pre-Apollo theories are wrong, because if either the fission theory or the double-planet theory is correct, the composition of the Earth and the Moon should be virtually the same.

According to the new theory, which was only widely accepted a decade after the last Apollo mission[5], soon after the Earth was formed it was hit by a large object. This was much larger than the object that caused the KT extinction and was probably about the size of Mars. The collision shattered a large part of the Earth, debris was thrown into space, and the Moon was formed as the debris coalesced. The theory explains why volatile substances are scarce in Moon rock – the heat of the collision boiled away most of the volatile material and the Moon was formed out of dry refractory stuff. Since the Earth's core remained relatively untouched by the impact, most of the material thrown into space came from the less dense mantle – and so the impact theory also explains nicely why the Earth is denser than the Moon and why the Earth has a core but the Moon does not. Computer simulations of such an impact produce something very like the Earth–Moon system, and since no other theory is now consistent with the evidence, it seems virtually certain that this is how the Moon was formed.

And so the characteristics of the Solar System today seem to be partly the result of gradual processes operating in a disk 4.6 billion years ago and partly the result of the roulette wheel of impacts. Chance also seems to have played a part in the origin of our species, because if the KT impact had not conveniently cleared out the dinosaurs it seems unlikely that Homo Sapiens would be here now. Much of the evidence for the importance of chance in the story of the Solar System has come from the analysis of the Moon rock. Apollo did cost a lot of money. The nineteen billion dollars might have been spent instead on feeding the hungry in the Third World and in solving social problems in the First World, although I still think this money would not have made much of a dent on these problems. However, Apollo was possibly the greatest ever human adventure – the first voyage of our species to another world. The results from Apollo have also been vital for understanding the origin and history of our planetary system.

If I could travel back in time thirty years, that is what I would say.

2. The Day the Solar System Lost a Planet

In the history of planetary exploration, March 13th, 1986, was a red-letter day. This was the day a spacecraft visited a comet for the first time.

Throughout most of our history comets have been much more important to the average human being than planets. Planets are merely points of light which can only be distinguished from stars because they move around the sky; they can be safely left to people who professionally have to worry about such things – astronomers and astrologers. A comet though is harder to ignore. When Comet Hale–Bopp was in the sky in 1997 I saw people who I knew had little interest in astronomy standing in the street staring at it in astonishment. It is hard not to believe that a comet, which appears from nowhere and streaks across the sky, has some significance for human affairs. For most of human history people have believed that comets presage disaster. Comet Halley, the most famous comet of all, appeared at the destruction of Jerusalem by the Emperor Titus and at the Battle of Hastings; and perhaps the comet gave the Jews and Saxons some comfort that their defeats were due not to inferior military tactics or to a chance arrow but to the inexorable workings of fate. It is only since astronomers have shown that comets, like planets, are prisoners of the law of gravity, and that the tail of a comet is simply gas which has evaporated from the comet's nucleus as it moves towards the Sun, that comets have lost their ability to terrify.

Although comets lost their aura some time in the seventeenth century, there remained one big mystery. The tail of a comet can stretch a long way across the sky, often several times the width of the Full Moon, but the solid object that produces the gas that forms the tail – the comet's nucleus – is tiny, far too small to observe with a telescope from the Earth. Although most scientists

FIGURE 2.1 Picture of Comet Halley on its return in 1986. Credit: photograph taken by W. Liller, NSSDC/NASA

assumed that the nucleus is a mixture of rock and ice but mostly ice – a "dirty snowball" – until the mid-1980s nobody had ever seen the nucleus of a comet.

At the beginning of this decade, space scientists around the world realized that they had a great opportunity because Comet Halley (Figure 2.1), which travels around the Sun every 76 years, was due to come close to the Sun again in 1986. NASA, the European Space Agency (ESA) and the Japanese and Russian space agencies all planned to send missions to visit the comet. The original plan was for the Japanese and the Russian spacecraft to make long-distance measurements of the comet, while a joint US–European spacecraft would travel close to the nucleus. However, the Americans dropped out because of lack of money, and the final mission to the heart of Comet Halley was a purely European affair.

I do not have particularly warm memories of Comet Halley myself, because it marked the nadir in my career as an astronomer. As Giotto, which was named after the medieval Italian painter who painted the comet on one of its other returns, headed out into space for its history-making encounter, my own career seemed to have stalled irrevocably. I had just completed a Ph.D. at Cambridge University and I was at that awkward stage in the career of every astronomer of trying to get my first "postdoc," a three-year contract to do fulltime research. While I was still trying to finish my Ph.D., every few days a rejection letter from some prestigious university around the world would appear in my mailbox. As rejection letter followed rejection letter, I lowered my sights. I started to apply for jobs at less fashionable universities, and I even began to consider jobs outside astronomy. As a last throw before I applied for a job in the real world, I applied for a postdoc at the University of Kent, which would have involved analyzing observations of Comet Halley. My Ph.D. had been in cosmology and so there was the minor difficulty that I actually knew virtually nothing about comets. However, I made this completely clear in my application, and I reasoned that either they would simply throw my application in the bin – in which case little time had been wasted – or, in the unlikely situation that nobody with any knowledge of comets had applied for the job, I might get it. A few weeks later, I got a phone call from the University of Kent asking me to come down for an interview that day. I spent an anxious four hours on the train travelling to Canterbury and an hour being interviewed, during which my lack of knowledge of comets was uncovered in humiliating detail. When I got back to Cambridge, I started to apply for jobs outside astronomy.

Giotto was the first European deep-space mission, and it was a challenging one to start with. The big problem was the possibility of the spacecraft being damaged by the gas and solid fragments streaming away from the nucleus; this material is moving so fast that even a one-gram fragment of ice packs about the same punch as a bag of potatoes falling on ones head from the top of a skyscraper[6]. Because of their concern over the spacecraft's safety, the ESA scientists decided that the closest Giotto would get to the nucleus would be 500 kilometers.

Space missions are always nerve-jangling affairs, but because of the possibility of the spacecraft being damaged by material from the comet, the Comet Halley flyby was more nerve-jangling than most. As the spacecraft approached the comet's nucleus, all the scientists who had designed the scientific instruments on the spacecraft gathered around the TV screens at the European Space Operations Centre in Darmstadt watching the raw images from the Halley Multicolor Camera build up on the screens. Gerhard

Schwehm, who was then the deputy project scientist, recalled: "It was a very exciting time but also a time of tension because we didn't know whether the spacecraft would operate properly. It was vital that it did because the actual encounter lasted only a few hours and there was no time for recovery if anything went wrong."[7]

The first of 12,000 impacts occurred 122 minutes before closest approach. Images continued to appear on the screens as Giotto closed to a distance of 1400 kilometers, but the rate of impacts increased sharply as the spacecraft passed through a jet of material streaming from the nucleus. Finally, eight seconds before closest approach, and only 596 kilometers from the nucleus, a solid fragment with a mass of about one gram smashed into the spacecraft and sent it spinning. The TV screens went blank and the waiting scientists feared that this was the end. After a few seconds, though, a burst of information came through from the spacecraft. Over the next half an hour, sporadic bursts of information were received, and eventually, as the spacecraft started moving away from the nucleus, full contact was restored. Some of the sensors were permanently damaged by the impacts, but enough remained for Giotto to visit a second comet successfully six years later, and the information Giotto collected about Comet Halley has permanently changed our view of comets.

For a start, the nucleus of Comet Halley does not look like a dirty snowball. It actually looks more like a blackened peanut than a snowball (Figure 2.2). The surface of the nucleus is not smooth like the surface of a snowball but, like other astronomical bodies, contains hills and depressions. The dark surface, which is darker than coal, is probably explained by a thick layer of dust. There must however be a lot of ice under this dark surface, because Giotto's instruments showed that most of the gas streaming from the surface is water vapor. The gas is not streaming uniformly off the surface. Giotto's images showed that much of the surface is inactive and that jets of gas come from only a few spots. The probable explanation of these jets are pockets of ice, which will produce sudden jets of gas when the intensity of the Sun's heat reaches a high enough level – a possibility which makes the neighborhood of a comet's nucleus an even more dangerous place.

In Chapter 1, I told the story of the origin of the Solar System, at least as far as we have been able to piece it together today. I left

FIGURE 2.2 Image taken by Giotto of the nucleus of Comet Halley. Credit: ESA/Giotto

out one important part of this story because it is worth a chapter in its own right. In this chapter, I want to tell the story of the origin of the comets. As is true of many of the stories in this book, this story is a mixture of science and history.

The story starts with the scientific fact that there are two types of comets. Comet Halley is a short-period comet. Although this comet appeared to the Jews, to the Saxons, to Giotto and to many others down the ages, everybody assumed that each time it was a different comet until Edmund Halley used Newton's newly minted theory of gravity to show that it was actually the same object returning. Comet Halley orbits the Sun in a highly elliptical orbit, and every 76 years we see it when it gets close enough to the Sun for the heat to melt the ice and for the tail to form. Although 76 years is long in human terms, it is shorter than the orbital periods of Uranus, Neptune and Pluto. Like planets, short-period comets are objects that are bound to the Solar System, and like planets they orbit the Sun in roughly the same plane.

Most of the comets that appear in the sky each year, however, are not short-period comets. Long-period comets really do seem to come from nowhere. Analysis of the orbits of these comets shows that long-period comets are new comets, comets that have never been in the Solar System before and that, once they leave it, will never return. Twenty of these comets are seen every year, which means that if the rate has stayed the same since the formation of the Solar System 4600 million years ago, roughly one hundred billion different comets must have visited the inner Solar System in that time. This calculation led the Dutch astronomer Jan Oort to suggest in 1950 that the Solar System must be at the center of a cloud of dirty icebergs. Oort's idea was that, every so often, the gravitational force exerted by a nearby star would dislodge one of these icebergs, which would then plummet into the Solar System to become a comet. If it exists, the *Oort Cloud* is immense, both in scale and in the number of icebergs it contains. Pluto is about 39 times further from the Sun than the Earth. The distance from the Earth to the Sun is defined as one Astronomical Unit (AU), so Pluto is 39 AU from the Sun. But the radius of the Oort Cloud is 50,000 AU, almost half way to the nearest star. Estimates of the number of icebergs it contains range from one hundred billion to one million billion. The Oort Cloud is a strange thing. Most astronomers think it must be there – the long-period comets have to come from somewhere – but the icebergs are so small and so far from the Sun that nobody has been able to think of a way of observing it.

It used to be thought that the two families of comets must be related, and that the short-period comets are the rare long-period comets that *do* get trapped within the Solar System. The popular idea was that if a long-period comet gets close enough to one of the two biggest planets, Jupiter and Saturn, the gravitational force exerted by the planet would be enough to transform the comet's orbit into that of a short-period comet.

In the 1980s, it gradually became clear that this family connection is illusionary. The growth in the power of computers made it possible for the first time to construct computer models of the Solar System that could be used, like a clockwork model, to follow the paths of planets and comets millions of years forward into the future and backwards into the past. The models showed that the chance of a long-period comet being turned into a short-period comet is actually very small. They also revealed an even more fundamental problem. The long-period comets, since they come from a spherical cloud, bombard the Solar System from all angles; the short-period comets orbit the Sun in the same plane as the planets. The models showed that *if* a long-period comet were caught, it would almost certainly not end up orbiting the Sun in the same plane as the planets. By the end of the 1980s, the problem of the origin of the short-period comets was one of the major unsolved problems of planetary astronomy.

At this point, I have to pull myself up short. One of the worst sins for even an amateur science writer is to gloss over the chaotic messy way that science often happens. If I were not too worried about journalistic accuracy and just wanted to tell a neat story, I would write that planetary scientists worried about this problem for many years, and eventually Dave Jewitt and Jane Luu at the University of Hawaii designed a clever observing program which produced the solution. As far as I can judge, since I do not work in this research field myself, this was not the way it really happened. Although by the end of the 1980s there were some scientific papers claiming that the popular idea could not be correct, one does not usually start to worry too much about something like this (in my own research field at least) until there is much stronger evidence; one paper, a few papers even by well-respected scientists, can so easily be wrong. My hunch is that Jewitt and Lu were trying to do something much simpler, and that it was only by the weird synchronicity that is common in the history of science that they produced a solution for the problem at virtually the same time that people realized there was a major problem – or rather, once Jewitt and Luu had the solution, people looked back at the papers from the 1980s and realized that, yes, there had been a major problem*.

---

* I have left this section as I wrote it originally. Dave Jewitt subsequently read this chapter and confirmed that my guesses were largely correct. He and Luu were *not* trying to discover the origin of the short-period comets but instead, as is described later in this chapter, do something rather simpler.

This sin's attraction is that it would make it possible to tell the story with a single narrative line. However, if I am to describe Jewitt's and Luu's discovery in the way it actually happened, I must now jump back in time two hundred years and begin a second storyline.

The simplest way of becoming famous as an astronomer is to discover a planet. Of course, most of the planets have been known forever. Mercury, Venus, Mars, Jupiter and Saturn are all bright enough to be seen with the naked eye and, because they move relative to the fixed stars, have been known ever since humans first looked at the sky. However, in the eighteenth century an astronomer did discover a new planet, and in the process made himself a celebrity.

The person whose work led to the first expansion of the Solar System was a monomaniac with two good ideas. Until he was forty-three years old, William Herschel was a music teacher and only an amateur astronomer, but his obsession with astronomy meant that he spent more time, money and energy on his hobby than most professionals. When grinding a mirror for a telescope, so the story goes, he would spend ten hours at a stretch bent over the grindstone, taking no time off for meals but being fed morsels of food by his sister Caroline while he worked. His first good idea was that he realized the key to seeing fainter objects, and so to looking deeper into the Universe, is to build telescopes with larger mirrors. This started the progression to bigger and bigger mirrors that has culminated in the 10-meter mirror of the Keck Telescope (Chapter 7). However, unlike the Keck Telescope, designed and built by a multitude of scientists, engineers, and technicians, Herschel constructed all his telescopes himself, and his elegant Georgian house in the English spa-town of Bath fulfilled the combined functions of home, observatory, and telescope factory*. Herschel's second good idea was that he realized progress in astronomy does not just come from observing known objects but

---

* An interesting literary link is that this was the period of Jane Austen's novels, in which Bath is a frequent location. Music teachers are also often mentioned, and I like to imagine Herschel teaching music to Miss Austen's fashionable young ladies.

also from carrying out systematic surveys of the sky. His method of surveying, which he called "sweeping," was each night to choose a strip of sky about two degrees wide and then go through it twice looking for new objects. In the course of his career, he managed to survey the whole sky in this way five times; each time, because he was using a new telescope with more sensitivity than any that had gone before, he found hundreds of new star clusters, nebulae, and double stars.

The meridional moment of Herschel's life occurred one night in his back garden in Bath. While sweeping the sky with a new telescope, he noticed a star that appeared to have changed its position relative to the other stars. He had discovered a new planet, the first planet discovered in modern times. The discovery changed his life. From being an almost unknown amateur – the Astronomer Royal did not know how to spell his name – he became a celebrity. He spent three months being cosseted by fashionable hostesses in the drawing rooms of London society (a complete waste of time, he wrote to Caroline) and a pension from George III allowed him to give up teaching music.

Before the new planet had travelled around the Sun once – Uranus takes eighty-four years to travel around the Sun – another planet had been discovered. The path taken by a planet is mostly governed by the gravitational force of the Sun, but the gravitational forces of the other planets also have a small effect. In the years after Uranus was discovered, astronomers noticed that it was not quite following its predicted path, even when the gravitational forces of the Sun *and* all the planets were taken into account. Two scientists, Urbain Leverrier of France and John Couch Adams of England, independently realized that there must be a planet outside the orbit of Uranus disturbing its orbit. Using the mathematics of Newton's theory of gravity, they predicted where this planet must be. In 1841, close to the predicted position, the German astronomer Johann Gottfried Galle discovered the planet Neptune.

If chance led to the discovery of the seventh planet and mathematics to the discovery of the eighth, sheer boring, bottom-numbing hard work led to the discovery of the ninth. Even after Neptune's gravitational effect was taken into account, Uranus did not follow the predicted path exactly. However, the discrepancy between its predicted and its actual path was so tiny, only 2% of the one that had been present before the discovery of Neptune, that it was not clear whether this was significant. Nevertheless, every now and then, an astronomer would suggest that the discrepancy might be due to the gravitational effect of a planet beyond Neptune, and make a half-hearted attempt to look for the planet. However, the small size of the discrepancy meant that it was impossible to calculate precisely where the planet would be in the sky, and it also meant that the planet must be very faint – and for almost a century the attempts failed. One of the places at which searches for a ninth planet had been intermittently underway was Lowell Observatory in Arizona. By 1929, it had become clear to the director of the observatory, Vesto Slipher, that the only way another planet would ever be discovered would be by a careful survey of the whole sky. The person he hired to carry out this survey was a 22-year-old farm boy called Clyde Tombaugh.

Tombaugh did not have the years of training of the professional astronomer (he did not even have a university degree), but a professional astronomer would never have had the patience for the survey planned by Slipher. For this survey, the daily routine and monotony of farm work were the ideal preparation. At the time, the best way of spotting a planet was still to use the human eye. Tombaugh's tool was the blink comparator, which is a device for displaying in rapid alternation two photographic plates of the same area of the sky taken at different times; any object that has moved in the interval jumps out to the eye. Over a period of 14 years, Tombaugh used the blink comparator to look at plates covering 70% of the sky. He estimated that by the end of this period he had stared at images of ninety million astronomical objects[8]. Of these ninety million, all except for 3970 showed no movement and were therefore stars. Tombaugh used one of Kepler's laws of planetary motion (see below) to show that of the remaining 3970, 3969 were moving so rapidly that they must be asteroids, small objects orbiting the Sun between the orbits of Mars and Jupiter. The remaining object, the last of ninety million, was the planet Pluto.

Pluto has always been the planetary misfit. It is on the outskirts of the Solar System, but it is not a gas giant like Jupiter, Saturn, Uranus and Neptune. Instead it is a small solid lump like the inner planets (it is actually smaller than the Moon). While the other eight planets move around the Sun in orbits that are almost circles, Pluto has a highly eccentric orbit. Although it is currently the most distant planet in the Solar System, at other times in its 248-year orbit around the Sun it will lose the mantle of most-distant planet to Neptune.

Herschel, Adams, Leverrier, Galle, Tombaugh – by their discovery of new planets all of them have entered the pantheon of astronomy. Given the lure of a kind of immortality, it is not surprising that since the discovery of Pluto searches for new planets have never entirely stopped. Until the end of the 1980s, the most thorough search for Planet X* was carried out by Charles Kowal at Mount Palomar. Using a 48-inch telescope, with fourteen times the light-gathering power of the 13-inch telescope used in Tombaugh's survey, Kowal spent seven years searching for a tenth planet. He found five new comets and 15 asteroids – but no new planet. Kowal's conclusion, after seven years staring through a blink comparator, was that further searches for extra planets were just not feasible; they were too time-consuming and exhausting and the lure of your name in the history books just wasn't worth it.

Nevertheless, at the end of the 1980s, Dave Jewitt and Jane Luu, who worked at the Institute for Astronomy (IFA) in Hawaii, decided to try again. By a sheer fluke, I also happened to be working at the IFA at the time, although unfortunately I never met them while I was working there. After being rejected, it seemed, by virtually every third-rate astronomy group in the world, I had finally succeeded, for no obvious reason, in landing a postdoc at one of the very best places, the IFA. At the end of 1985, I left England behind and flew out for a three-year postdoc in Honolulu.

Honolulu extends about twenty miles along the southern side of Oahu, the second-largest island in the Hawaiian chain, on the narrow coastal plain between the ocean and the mountains. The institute is in one of the many valleys that cut back into the mountains. In Hawaii, weather is a matter of geography rather than time. On the beach the sun is always shining; in the mountains, only a few miles away, it is usually raining. From my office window half

---

* X is the symbol for the unknown as well as being conveniently, given the number of planets in the Solar System, the Latin symbol for ten.

way up Manoa Valley I could usually see both the sun and rain, and most afternoons there would be a rainbow arching across from one side of the valley to the other. Many houses in the valley have been owned by the same families for generations, and with its ramshackle houses, winding lanes, and the sound of cockerels, the valley feels a thousand miles in spirit from the hotels only two miles away on the coast. The institute itself is a white two-storied building, designed in the Hawaiian style to let in air and light. All in all, it was a beautiful place to spend three years.

However, the reason the institute is a mecca for observers is not the beautiful scenery, but something that cannot be seen from the windows of the building. At the time, the world's best observing site was Mauna Kea, an extinct volcano on the Island of Hawaii, the biggest island in the chain – the "Big Island." For an observer, the Earth's atmosphere is a problem; it distorts the images of faint stars and galaxies and absorbs the radiation one is painstakingly trying to detect. At a height of 14,000 feet above sea level, the Mauna Kea Observatory is already above 40 per cent of the atmosphere, making it possible to obtain pictures of the Universe of exceptional clarity. Although there are now other sites that have as good, or arguably better, observing characteristics*, Mauna Kea is still the biggest of the top observatories. There are now 13 telescopes on the summit and almost all observers have observed there at some time or other. When the quality of the site was first realized, the government of the State of Hawaii decided they did not want foreign astronomers simply jetting in and jetting out without contributing anything to the people of the State. They took the decision that anyone who wished to build a telescope on Mauna Kea would have to give at least ten per cent of the observing time on the telescope to the Institute for Astronomy. I do not know what brought Jane Luu and Dave Jewitt to Hawaii, but I assume it was the same thing that brought me there: the lure of observing time on the big telescopes on Mauna Kea.

In the late 1980s, Jewitt and Luu started a new search for objects outside the orbit of Pluto. In the interval since Kowal's

---

* The Atacama Desert in Chile, which is at a height of 18,000 feet, and Antarctica are better for kinds of astronomy for which the dryness of the atmosphere is critical.

search, there had been one big change in astronomy: the development of CCD cameras. CCD stands for *charge-coupled detector*, which is not terribly informative, but essentially a CCD camera is an array of tiny light-sensitive detectors. The digital cameras that can now be bought for less than one hundred dollars in any mall are CCD cameras, but in the 1980s one property of CCD cameras, which is not realized by someone using a digital camera to take a picture of the dog or grandma, produced a revolution in astronomy: CCD cameras are supremely sensitive. Light consists of tiny particles called *photons*. Whereas a photographic plate will detect only about one out of every 20 photons from a faint galaxy, thus wasting 19 of the photons, a CCD camera will typically detect 14 out of the 20 photons. This property is not important for everyday photographers who are swamped with photons, but it is critical for astronomers for whom every photon is precious. CCD cameras however do have one downside for astronomers. They have a very small field of view, which means it is only possible to survey very small areas of sky.

Jewitt and Luu's search was therefore rather different from those of Tombaugh and Kowal. They knew from the earlier surveys that there could not be any really large object in the outer Solar System. Instead, they decided to use the ability of the new CCD cameras to survey small areas of sky with great sensitivity to look for small objects in the outer Solar System. Their project was not carried out in a complete theoretical vacuum. As I will describe below, there had been suggestions over fifty years earlier that there might be small objects beyond the orbit of Pluto, and that these small objects might be connected to the origin of the short-period comets. I do not know whether these suggestions were strong motives for Jewitt and Luu's survey. My suspicion is that their primary motive was the basic observer's desire: to look out into the Universe to see what is there*. Their basic method, however, was exactly the same as Kowal's and Tombaugh's: take

* Dave Jewitt has confirmed that they were unaware of this earlier work. Their primary motivation, according to Jewitt, was the suspicion that since there are many small objects in the inner Solar System, there might well be some hiding in the Solar System's outskirts – something which had not been possible to test before the advent of CCDs.

two images of a piece of sky at different times and look for objects that have moved relative to the fixed stars. Also like Tombaugh's and Kowal's surveys, their survey lasted many years without success. In Jane Luu's words:

> For five years, we continued the search with only negative results. But the technology available to us was improving so rapidly that it was easy to maintain enthusiasm (if not funds) in the continuing hunt for our elusive quarry. On August 30, 1992, we were taking the third of a four-exposure sequence while blinking the first two images on a computer. We noticed that the position of one faint "star" appeared to move slightly between successive frames. We both fell silent. The motion was quite subtle, but it appeared definite. When we compared the first two images with the third, we realised that we had indeed found something out of the ordinary. Its slow motion across the sky indicated that the newly discovered object could be travelling beyond even the outer reaches of Pluto's distant orbit. Still we were suspicious that the mysterious object might be a near-Earth asteroid moving in parallel with the Earth (which might also cause a slow apparent motion). But further measurements ruled out that possibility[9].

Figure 2.3 shows the four images. There is a diagonal streak on three of the images. This is an object that is moving so fast that it has moved significantly even within a single exposure. In the early seventeenth century, the astronomer Johannes Kepler discovered a law that relates the time a planet takes to go round the Sun to its distance from the Sun*. The law predicts correctly that the more distant planets from the Sun move across the sky more slowly. Because of Kepler's law, Jewitt and Luu realized the streak must be an asteroid. Asteroids range in size from the largest asteroid, Ceres, which has a diameter of about 1000 km, down to objects the size of pebbles. There are thousands of asteroids known and Jewitt and Luu had little interest in them. The object that did

---

* Kepler's third law is that the square of the orbital period of a planet (the time it takes to go round the Sun) is equal to the cube of its distance from the Sun.

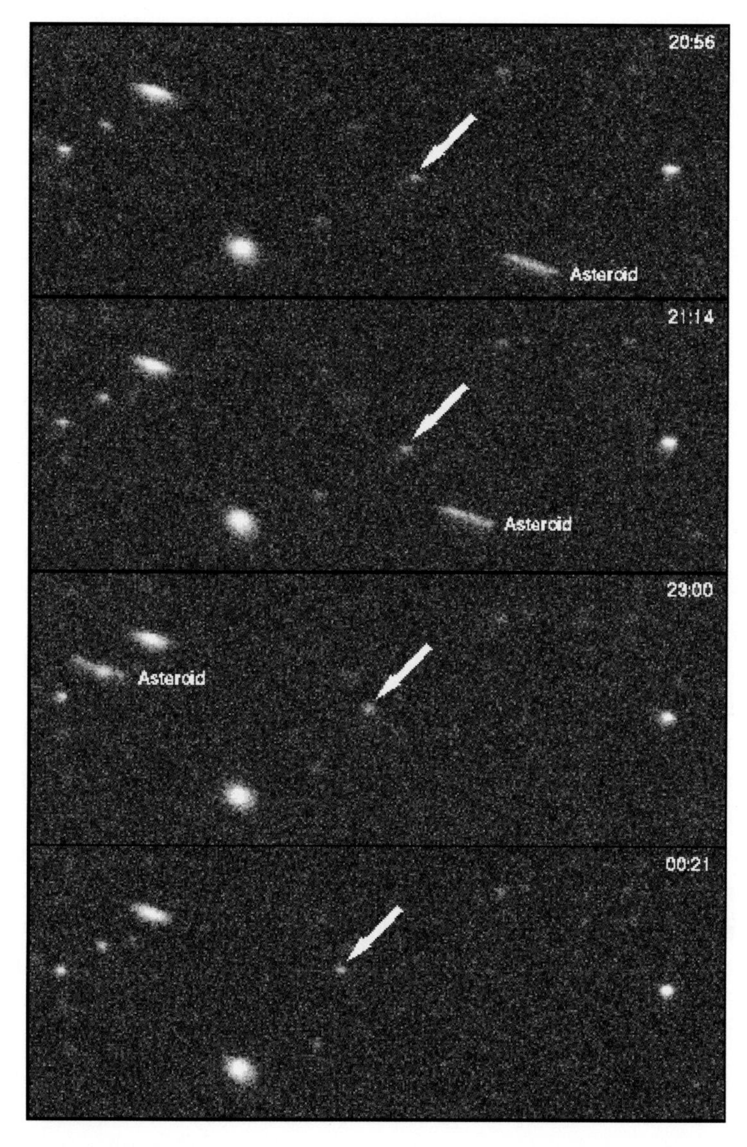

FIGURE 2.3 The images in which a new part of the Solar System was discovered. The images were taken one after another. The positions of the stars are the same in the four images; the streak is a fast-moving asteroid; the object whose position is slowly changing from image to image is the object discovered by Luu and Jewitt. Credit: David Jewitt

excite them is the object marked by an arrow. It *is* moving, but this movement can only be seen from its slightly different positions in the different images. Because of its slow speed, Jewitt and Luu suspected it might be a long way from the Sun.

After a good day's sleep down at Hale Pohaku, the astronomers' hostel, Jewitt and Luu returned to the summit the following night. One possible alternative explanation was that the object is an asteroid which happens to be on an unusual orbit taking it towards the Earth. By careful measurements of the position of the object on this night and the next night, Jewitt and Luu were able to rule out this possibility. They were able to show from Kepler's law and their measurements that the object is orbiting the Sun at a distance of 40 AU, slightly beyond the orbit of Pluto. The brightness of any object in the Solar System depends on its distances from the Earth and the Sun and on its size, which determines how much of the Sun's light it intercepts and reflects towards us. Once Jewitt and Luu had an estimate of the object's distance, they were able to calculate from its brightness that the object had a diameter of about 250 km.

They knew they had made an important discovery. The first thing to do when you make a discovery like this is to send off a telegram to the International Astronomical Union. This registers your discovery and it also alerts astronomers world-wide that there is something interesting in the sky. This is particularly important for something transient like a comet or a supernova (Chapter 4) because it is important to train as many telescopes as possible on the object before it vanishes from view (and also to stake your claim before anyone else notices it). Once Jewitt and Luu had assured themselves that the object was in the outer Solar System, they fired off an IAU telegram. They wanted to call the object "Smiley" after John Le Carré's fictional spy (the object had successfully eluded astronomers for so long and *Tinker, Taylor, Soldier, Spy* had recently been shown on public television). The rather more conservative IAU gave it a less literary name based on its date of discovery: 1992_QB1.

1992_QB1 or Smiley has a diameter about one tenth that of Pluto, making it, if it is a planet, much the smallest planet in the Solar System. However, Jewitt and Luu continued their search and in March 1993 they discovered a second trans-Neptunian object (outside the orbit of Neptune). By the end of 1993 they had discovered an additional four objects and, as I write this, there are over 800 trans-Neptunian objects known. The total number must be much greater because only a very small part of the sky has been searched; the most recent estimate is that beyond the orbit of Neptune there are approximately 100,000 objects with diameters greater than 100 km. Jewitt and Luu had not discovered a new planet, but instead they had discovered an entirely new component of the Solar System, a belt of small objects orbiting the Sun outside the orbit of Neptune – a discovery that should definitely earn them a place in the astronomical pantheon.

As I hinted above, the existence of this belt was not actually completely unsuspected. Over forty years before Smiley was discovered, two scientists, an Irish amateur Kenneth Essex Edgeworth and an American professional Gerard P. Kuiper, suggested that a belt of objects beyond the orbit of Neptune might be the source of the short-period comets. Kuiper's argument was that there was no reason why the original solar disk should have cut off abruptly at the position of the last two planets; beyond that point the disk might have been too tenuous for planets to form but still dense enough for the formation of smaller objects. He suggested that the temperature of any objects that had formed there would be so low that they would be composed mostly of ice and various other frozen gases – very similar to the nuclei of comets*.

Luu and Jewitt's discovery immediately made the Edgeworth–Kuiper (EK) Belt the prime suspect as the source of the short-period comets. The case against the EK Belt was proved by a smoking gun. Kowal, during his vain search for a tenth planet,

---

* In a miscarriage of justice, the belt is usually called the 'Kuiper Belt'. Edgeworth was an amateur scientist who did not belong to any university or observatory, in the Victorian tradition of the gentleman-amateur. The belt is usually called after Kuiper, who was part of the astronomical establishment, although Edgeworth suggested its existence two years before Kuiper. Most professionals do recognise the priority of Edgeworth, but the weight of custom has meant that it is generally referred to by astronomers, both in private and in print, as the 'Kuiper Belt'.

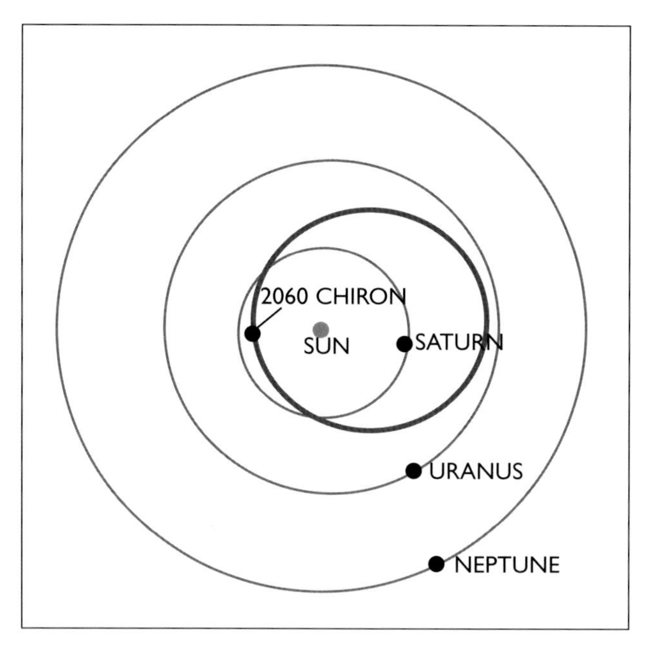

FIGURE 2.4 Orbit of 2060 Chiron.

had discovered an unusual asteroid, 2060 Chiron (Figure 2.4). This object is unusual because its orbit is not confined within the Asteroid Belt but instead crosses the orbit of Saturn and extends out almost to the orbit of Uranus. Large planet-crossing orbits like this are not stable; within a few million years the gravitational effect of the giant planets will either hurl Chiron out of the Solar System or send it into an orbit that takes it much closer to the Sun. Astronomers have recently discovered that Chiron is surrounded by a thin layer of gas. Chiron is clearly a comet, with the gas produced by the evaporation of ice from its surface, and so, metaphorically at least, it is smoking. At the moment, Chiron is in the outer Solar System, where the sunlight is very weak, and so it is a distinctly unimpressive comet. In a few million years, however, if the roulette wheel of gravity takes Chiron into the inner Solar System, our distant descendants may see it blaze across the night sky. Since the discovery of Chiron, several other objects have been discovered with similar orbits. One member of this group, known as the Centaurs, is a particularly strange object. Until 1992, this object, 5145 Pholus, had the reddest colors of any object in the

Solar System. Smiley however has almost exactly the same colors. This docket of evidence makes it almost certain that the Centaurs were once in the EK Belt, that they were dislodged by the gravitational force of Neptune, and they are now on their way to becoming short-period comets. An open-and-shut case.

The discovery of the EK Belt is also important because its existence is evidence of a process that may have completely transformed the Solar System shortly after its birth. This evidence is that it is quite difficult in the standard model of the origin of the Solar System, which I described in the last chapter, to see how it was possible for the objects in the EK Belt and in the Oort Cloud to have been formed at all.

The standard model, remember, goes like this. Before the birth of the Solar System, there was a large cloud of gas. As a result of gravity, the cloud collapsed to form a rotating disk. The Sun formed at the center of the disk; the disk cooled; chemical compounds with high melting points began to freeze; solid particles formed within the disk; the solid particles gradually stuck together, eventually forming the planets. As I described in the last chapter, the most uncertain step is the last one – from tiny solid particles to planets. However, it seems likely that an important intervening step would have been the formation of *planetesimals*, bodies perhaps about 100 kilometers in size. The Solar System was probably filled at one time with billions of these objects, and it was almost certainly the gravitational attraction between them that led to the formation of at least the inner planets. Their expected size is similar to the size of Smiley, and in the standard model a natural explanation of both the EK Belt and the Asteroid Belt is that these contain planetesimals that somehow escaped being subsumed into planets.

The problem with this explanation, however, is that it is hard to see how the EK objects could have formed at their present positions. Although we still do not understand fully the step between the tiny particles that first formed in the disk and planetesimals (Chapter 1), it is clear that the rate at which planetesimals were formed would have depended on the density of the disk. In dense parts of the disk, a particle would have rapidly encountered other particles and would have grown like a snowball rolling down a hill; in parts of the disk with a low density, a particle would have rarely encountered other particles, and planetesimals would have been formed very slowly or not at all. The density of the solar disk 4.5 billion years ago in the vicinity of the present-day EK Belt can actually be estimated in a remarkably simple way. First, estimate the total amount of material that is contained today in all the objects in the EK Belt (this is admittedly still very uncertain). In your mind's eye, then imagine all this material smeared out in a ring around the Sun – this ring represents that part of the disk out of which the EK objects were formed 4.5 billion years ago. From the estimate of the total amount of material today in the EK objects, the density of the material in the ring can then be estimated. The disturbing result is that the estimated density is so low that it seems impossible that any planetesimals would ever have been formed. This is an even worse problem for objects in the Oort Cloud.

One possible solution is that the EK Belt used to contain much more material, and so the estimate above is too low. Collisions between objects in the belt – there are so many objects that collisions will be quite frequent – are definitely gradually removing material. The debris from these collisions has probably been discovered. The first human interstellar spacecraft are the Pioneer 10 and 11 spacecraft, which were launched in the 1970s and have now travelled well beyond the EK Belt, and so are the first spacecraft to have left the Solar System. Back in the 1980s, when both spacecraft were still within the orbit of Uranus, they detected a shower of tiny particles drifting in towards the Sun, which is probably debris from collisions in the belt. Nevertheless, it seems unlikely that enough material has been removed from the belt for this to be the solution to the problem. A more likely solution is that the objects in the EK Belt and in the Oort Cloud were not formed where we see them today.

Even after the planets were formed, the Solar System probably still have contained billions of planetesimals that had not been incorporated in any planet. Most of their orbits would have been unstable because of the gravitational effect of the planets. The fate of each planetesimal would have been a matter of chance. Some of them would have been consumed by the Sun or have collided with the planets and moons, explaining the scarred face of our Moon (Chapter 1). But the most likely fate would have been that sooner or later the planetesimal would have strayed too close to one of the giant planets. If it had encountered Jupiter, for example, the planet's huge gravitational force would either have hurled it out of the Solar System completely or sent it into a highly elliptical orbit extending into the outer Solar System. This orbit would not have been stable either, because it would have crossed the orbits of one or more of Saturn, Uranus or Neptune – and the planetesimal would have undergone additional encounters with planets, each time either being hurled into interstellar space or sent into an elliptical orbit taking it further away from the Sun. Thus it seems quite plausible that the objects in the EK Belt and in the Oort Cloud did not form where we see them today, but were born in the inner Solar System and have since emigrated.

This is a plausible argument, but there is also now some direct evidence for it. We may actually have seen some planetesimals on the voyage out from the inner Solar System, the last vestige of the vast tide of emigration that occurred four and a half billion years ago.

Most of the EK objects discovered during the last decade are in the "classical EK Belt" (in astronomy anything discovered over ten years ago tends to be referred to as "classical"). This consists of objects in roughly circular orbits with distances from the Sun ranging from 40 AU to 50 AU (Figure 2.5). However, some of the EK objects are in the so-called "scattered disk." These are objects with elliptical orbits that take them closer to the Sun than the objects in the classical belt, down to close to the orbit of Neptune, but also much further away from the Sun than the objects in the classical belt. The technical astronomical terms for the points on an orbit closest and furthest from the Sun are the perihelion and the aphelion; the objects in the scattered disk have perihelia between 30 and 38 AU but aphelia that may be as much as 100 to 3000 AU. Although there are many fewer objects known in the scattered disk than in the classical belt, the true populations may be actually quite similar, because it is only possible to detect an object in the scattered disk when it is close to its perihelion. A highly elliptical orbit with a perihelion close to a giant planet is exactly the orbit expected for a planetesimal that has wandered too close to the planet. Thus it seems likely that the objects in the scattered disk are planetesimals which have recently encountered

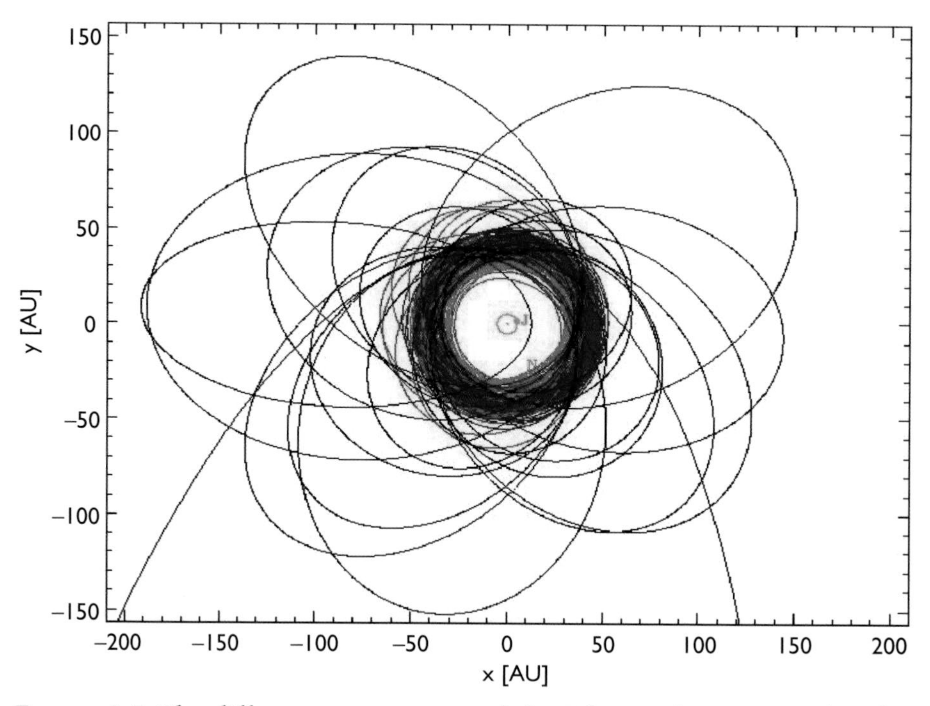

FIGURE 2.5 The different components of the Edgeworth–Kuiper Belt. The lines show the orbits of individual EK objects. The dense band of circular orbits in the center forms the classical EK Belt. The highly elliptical orbits are those of objects in the "scattered disk." Credit: David Jewitt

Neptune and are now embarked on the voyage out of the Solar System.

The scattered disk is almost certainly only a temporary staging post – an Ellis Island – for planetesimals emigrating from the inner Solar System. The stay of a planetesimal in the scattered disk can only be temporary because its orbit is still not completely stable*. Each time one of the objects in the scattered disk comes back close to the orbit of Neptune there is the possibility of additional changes to its orbit. These changes may take it back into the inner Solar System (the scattered disk is now thought to be the most likely source of the short-period comets), thus returning the planetesimal, to torture the metaphor, to its country of origin, or it may hurl it away from the Solar System completely.

* Temporary in the context of the age of the Solar System. The objects currently in the scattered belt may still survive for one billion years.

Today most of the planetesimals have moved on. Back at the time of the formation of the Solar System, this holding pen may have contained over a million times as many objects. At this time, the Sun would have been surrounded by many other newly formed stars (Chapter 5) and so there would have been stars much closer to the Sun than any today. Objects in the scattered disk move on elliptical orbits that take them far from the Sun. It is possible that at this time the gravitational attraction of nearby stars was enough to move many of the planetesimals from the scattered disk into a spherical cloud surrounding the Solar System – the Oort Cloud.

It may seem from what I have just written that astronomers now understand the origin of the EK Belt and the Oort Cloud. This is far from the truth. For a start, how was the classical EK Belt formed? The scattered disk is the obvious staging post for planetesimals moving from the inner Solar System to the Oort Cloud, but it is not obvious how planetesimals emigrating from the inner Solar System could have ended up in the highly circular orbits of the classical EK Belt. Then there is the problem created by a new family of EK objects that has been discovered within the last few years. The most famous member of this new family is the giant EK object Sedna which, with a diameter of between 1000 and 1500 km (the exact value is still uncertain), is not much smaller than Pluto, which has a diameter of only 2320 km. The thing that sets Sedna and the other four members of this new family apart is that they have highly eccentric orbits but large perihelia. Sedna, for example, has an aphelion of 500 AU but a perihelion of 76 AU, which means that even at its closest to the Sun it is over twice as far from the Sun as Neptune. Their distance from the Sun makes objects in this family very hard to find, and so this new family could easily contain more objects than the other two families combined. The problem posed by this new family, as with the objects in the classical belt, is that astronomers do not yet understand how planetesimals emigrating from the inner Solar System could end up on these orbits.

I want to finish this chapter with the ugly duckling of the Solar System, the chip of ice among the flamboyant gas giants: the planet Pluto. Pluto has a highly eccentric orbit that crosses the orbit of Neptune. Its orbit though is stable, because in the time it takes Neptune to orbit the Sun three times Pluto orbits the Sun twice, which keeps it safe from the gravitational effects of the larger planet. Objects with orbits like Pluto and Neptune are said to be in 3:2 orbital resonance. After the discovery of the EK Belt, it was soon discovered that some EK objects are in the same orbital resonance with Neptune – objects which were quickly given the name "plutinos." The discovery of the EK Belt and especially the discovery of plutinos gave rise to the uncomfortable suspicion that possibly Pluto is not really a planet but merely a rather large EK object. For almost a decade this suspicion remained for two reasons. Pluto was much bigger than the other EK objects, and Pluto was also unique because it has a moon, Charon.

The suspicion began to harden into something more definite in 2001 with the discovery of Sedna and other EK objects with diameters of over 1000 km. Moreover, by this time astronomers knew that Pluto was not unique in having a moon. There are actually nine EK objects which are now known to have tiny moons. Finally, in July 2005 it was announced that an EK object which is even bigger than Pluto had been discovered. When journalists heard about this object, which had the unglamorous name of 2003UB313, they raced off and wrote stories describing the discovery of the tenth planet – Planet X.

The status of Pluto was finally settled by a vote. On August 24th 2006, several thousand astronomers at a meeting of the International Astronomical Union in Prague voted to adopt a new definition of a planet. According to this new definition, to be a planet an object has to satisfy three conditions. It must orbit the Sun. It must be sufficiently large that its gravitational field pulls it into the shape of a sphere. And it must be much bigger than all the other objects in its orbital neighbourhood. Pluto, 2003UB313 (recently given the name Eris) and one of the asteroids, Ceres, satisfy the first two conditions, but they do not satisfy the third. The Solar System therefore now has eight planets. This definition was not adopted without a fight. A week earlier, a definition had been proposed which would have given the Solar System 12 planets, including Pluto, Eris, Ceres and even Pluto's moon, Charon. As a sop to Pluto's defenders, Pluto, Eris and Ceres – objects which satisfy the first two conditions but not the third – have been given a new title: dwarf planet. But this does not change the basic result.

The Solar System has lost a planet.

3. ET and the Exoplanets

One of the pleasures of living in the United States, which I did for four years, are the supermarket tabloids. These are found at every supermarket checkout, and every week, while standing in line to pay for my groceries, I used to catch up on the sex lives of the Hollywood stars and find out about the latest visit of extraterrestrials to the Land of the Free. I never actually used to *buy* one of the tabloids, you understand, but on the few occasions that I read the inside of one of them I found that all the important information was on the front cover anyway. The most downmarket of all the tabloids was the *Weekly World News*. This specialized in UFOs and alien abductions. A friend of mine claimed that it was actually surprisingly well written but, whatever its literary quality, I was never entirely convinced by its accounts of alien visits. The aliens always seemed to appear on a country dirt road to someone called Darlene or Jim Bob. If they wanted to contact us, why didn't they simply land on the White House lawn? And surely their technology was advanced enough, if they didn't want to contact us, that they could avoid being seen by rednecks on backcountry roads.

Ironically, the first attempt by scientists to contact extraterrestrials was made in exactly the kind of place that tended to feature in the *Weekly World News*. The Appalachian Mountains of West Virginia are one of the poorest and strangest parts of America, containing communities that were deposited in remote valleys by the tide of colonization two centuries ago and have remained isolated ever since. They are the home of hillbillies and the spiritual home of country music. The poverty is almost tangible. The narrow mountain valleys, a long way from Hollywood and Manhattan, are lined with trailers which often have black garbage bags in the windows rather than glass. For me, the drive through these dark valleys is always slightly frightening (my thoughts tend to veer uncomfortably to gun control legislation and to all those movies in which a nice young city couple get lost in the woods). Deep in the mountains, however, in the middle of this rural poverty, is one speck of high-tech affluence – Greenbank, the home of the National Radio Astronomy Observatory. In 1961, a young astronomer at the Greenbank observatory, Frank Drake, decided to use the 85-foot radio-telescope at the observatory to make the first search for radio signals from extraterrestrials.

I do not know whether or not Drake really expected to detect any signals[10], but he probably thought, as many scientists have after him, that the small chance of success was outweighed by the huge impact that any successful contact would have (although not necessarily a positive impact, as the *Weekly World News* or almost any science fiction book or movie shows). His reasoning was that radio waves are such a powerful means of communication that they would be the natural way an extraterrestrial civilization might try to contact us – much easier than visiting us by spaceship.

He was faced with two tough choices: which frequency to use and which star to use as a target. He decided after some thought that although there are billions of possible radio frequencies, there is one special frequency that extraterrestrials might use, hoping we would realize its significance. The most common element in the Universe is hydrogen, and because of a process within the hydrogen atom, which I do not have space to explain, hydrogen emits radio waves at a natural frequency of 1421 megahertz. Thus every cloud of gas in the Galaxy is broadcasting on this special frequency. Of course, any advanced civilization would know that this is a special frequency, and Drake decided this would be the obvious one to try[11]. The second decision was more difficult. Our Galaxy, which sprawls across the sky as the Milky Way, contains about three hundred billion stars. Which of these has planets around it? On which planets has life arisen? And on which of *these* planets has a technologically advanced civilization evolved? In 1961 nobody knew the answers to any of these questions.

Drake had to make an educated guess. Close stars were obviously better, because any radio signal would be easier to detect, but which nearby star should he choose? He decided to let himself be guided by the only example of a technologically advanced civilization he had. It has taken 4600 million years for our civilization to develop on the Earth. The Sun burns its nuclear fuel very slowly and there is enough fuel to last for another 5000 million years. But stars more massive than the Sun get through their fuel much faster; a star ten times more massive than the Sun will run out of fuel in only ten million years – much too short a time for life to evolve. A star less massive than the Sun will have a longer life, but at the price of providing hardly any warmth for surrounding planets. Drake made the conservative decision to choose two stars as targets that are very similar to the Sun: Epsilon Eridani and Tau Ceti. Both are about ten light years from the Earth.

Project Ozma (Drake took the name from Queen Ozma in the Oz books) lasted for two months. For two months, the telescope moved silently across the sky, tracking one or other of the stars. You know the result. If Drake had received a message from an extraterrestrial civilization, it would have been one of the most momentous events in human history, and everyone would have learned about it in kindergarten.

Drake could just have been unlucky. Even with a million extraterrestrial civilizations in the Galaxy, only one star in three hundred thousand would be circled by a life-bearing planet. He could also have been wrong about the frequency. There are billions of possible frequencies.

Of course, there is another obvious explanation: we may be alone in the Universe. We know that life can arise in the Universe because we are here, but it is remarkably difficult telling whether life is something which always arises given suitable conditions, or whether we are the beneficiary of a huge cosmic lottery win – perhaps the Sun is the only star with a life-bearing planet out of the three hundred billion stars in the Galaxy, or even the only one out of the ten thousand billion billion stars in the observable Universe (Chapter 9). Suppose that you have just won a real lottery. If you know how many people buy lottery tickets each week, you can calculate the probability of winning. But suppose you do not know anything about lotteries and have no way of finding out how many other people have bought lottery tickets. You will have no way of knowing whether winning a lottery is extremely unusual or whether it is something as routine as hearing your bank account is overdrawn.

We are actually not quite as ignorant as this. Frank Drake himself pointed out that it is possible to estimate the number of advanced civilizations with which we might communicate by multiplying together a large number of probabilities. First, take the number of stars in the Galaxy which have lives long enough for life to have evolved (if it is like Earth life) but which are not dwarf stars, which probably produce too little energy for life to have ever started. This is something we *do* know and is roughly one hundred billion. Then multiply this by the probability that a star has planets. If this is ten per cent, for example, the number of possible life-bearing stars is one tenth of one hundred billion, or ten billion. Multiply this by the probability that life actually starts on the planet; multiply this by the probability that life does not stick at the single-celled stage, which it did on Earth for three billion years, and multicellular life develops; multiply this by the probability that intelligent life evolves (this is not inevitable – the dinosaurs might still be around if the Earth had not been struck by an asteroid); and finally multiply this by the probability that the intelligent life produces the advanced technology necessary for communication. There is one other factor that has to be taken into account. If a technological civilization inevitably destroys itself – by nuclear war, a population explosion, global warming or by some other catastrophe – it is possible that during the long lifetime of the Galaxy millions of civilizations have flowered briefly, but that none has ever been in existence at the same time.

The problem with this kind of calculation is that none of these probabilities is known. Take the origin of life itself. Some of the basic ingredients of life, including amino acids, have now been found in many places in the Galaxy, but we still do not understand what animates these chemicals, what turns them into life. Given a stew of these ingredients, life might inevitably start, or life on Earth might be a huge rollover jackpot. Putting different, but perfectly plausible, probabilities into Drake's formula, I could argue either that there are millions of civilizations currently in the Galaxy or that there is only one.

One of the most interesting probabilities is that of a star having planets (Even if there is life nowhere else in the Universe, it would be nice to have somewhere to go if we ever develop interstellar space flight). As I described in the first chapter, it has always seemed quite likely that the formation of planets is a fairly routine business and, if the standard model is correct, that whenever a cloud of gas collapses to form a star a planetary system is also formed. However, the only convincing way to estimate this probability is to look for planetary systems around other stars, and until very recently this was not possible. (The existence of the Solar System is unfortunately no help – even if there is only one planetary system in the Universe, we must inevitably be living in it!) Fortunately, everything has now changed. In 1995 we knew of one planetary system. As I write, 156 planets have now been discovered around other stars, and the number is climbing all the time. The discovery of *exoplanets* is exciting for many reasons, but one of them is that it shows, since planetary systems are clearly fairly common, that the story of the origin of our planetary system that I told in the previous two chapters is almost certainly correct.

How can we look for planets around another star? The problem with the obvious answer – looking for one through a telescope – is that planets are very faint. A planet shines by the light reflected from its star and so a planet is always much fainter than the star around which it orbits.

Let us consider the actual problem of trying to detect planets around Alpha Centauri which, as the nearest star, should present the easiest problem in planet detection. Alpha Centauri is four light years from the Earth, which means it takes light four years to reach us (translated into regular units, this is the mind-spinning distance of 37,800,000,000,000 kilometers). Alpha Centauri is actually three stars, so close together that from the Earth they look like a single star. Alpha Centauri A and B are rather similar stars to the Sun. They are orbiting around a common point, with the two stars always on opposite sides of this point, which is called the *center-of-mass* (Figure 3.1). The distance between the two stars is approximately the same as the distance between the Sun and Neptune. The third star, Alpha Centauri C, is a tiny glow-worm of a star which radiates only about one ten thousandth of the energy of the Sun and orbits the other two stars at a distance of 50,000 AU, which is roughly the radius of the Oort Cloud (Chapter 2). About one half of all stars are in multiple star systems like this.

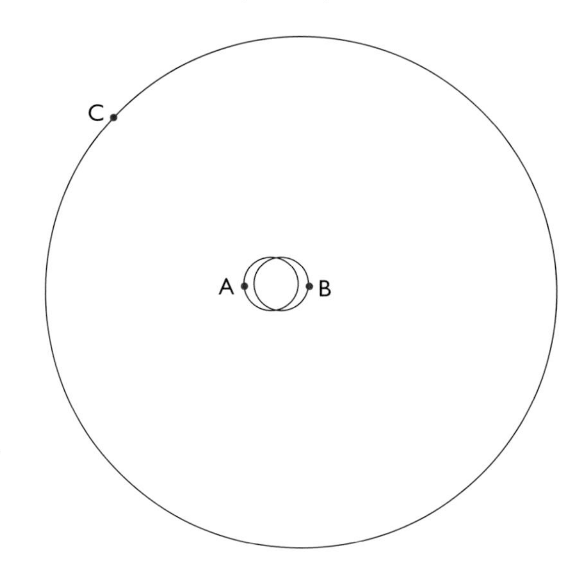

FIGURE 3.1 The Alpha Centauri system. The system consists of three stars, all orbiting a common center-of-mass. The circles show the orbits of the stars. Alpha Centauri A and B are separated by a distance approximately equal to the distance between the Sun and Neptune. The third star, Alpha Centauri C, is at a distance of 50,000 AU from the other two. The picture is not drawn to scale. If it were, the outer orbit would have a diameter of about 10 meters.

Astronomers used to think that a planetary orbit in a multiple star system would be unstable, but they now think that a stable orbit is possible if the orbit is close enough to one or other of the stars.

Life on a planet around one of these stars, however, would be different in some important ways from life in Kansas. Imagine that you are an extraterrestrial who lives on a planet around Alpha Centauri A*. The big difference you would immediately notice is that there is not one Sun in the sky but two. Alpha Centauri B is a thousand times fainter than Alpha Centauri A, but it is still two hundred times brighter than the Full Moon. For about half the year, both Suns are in the sky at the same time. Alpha Centauri B

* I am assuming that the planet is similar in most respects to the Earth, but that it is at a slightly larger distance from Alpha Centauri A than the Earth is from the Sun to compensate for the difference in temperature of the two stars.

is redder than Alpha Centauri A, and when the brighter sun passes behind a cloud the whole color of the landscape changes (artists on the planet work hard to capture the subtly changing hues). During this part of the year, night is the same as night on the Earth and the same constellations are visible. For the rest of the year, night is very different from night back in Kansas. Alpha Centauri B is now in the night sky. It is so bright that the constellations can not be seen at all and you can do the same things by night that you usually do by day. The light though is softer, and during this part of the year about twenty percent of the population, the dreamier artistic twenty percent, prefer to change their schedule so that they are awake when Alpha Centauri B is in the sky. Although in this season of the year there are some people who miss the starry sky, there is less traffic on the roads, murders and suicides are down, and psychologists claim that the mental health of the population is better.

Everyone writing about extraterrestrials always falls into the same trap: the anthropocentric fallacy. The aliens encountered on country dirt roads, as reported in the *Weekly World News*, are uncanny and scary, but they usually still have two legs, two arms, and two eyes. The producers of the TV show *Star Trek: The Next Generation* seem to have virtually given up the battle of trying to make actors look like aliens; the *Star Trek* aliens look remarkably like handsome Hollywood actors, with just a touch of latex and rubber to show their honest-to-God extraterrestrial origin. I have just jumped into the same trap, of course. I have assumed that the planet around Alpha Centauri A must be similar to the Earth and that the extraterrestrials have human concerns, such as art, psychology, murder and even traffic. However, to a large extent, the anthropocentric fallacy is an inevitable one. The only firm fact on which we can base our imagination of what extraterrestrials might be like is life as we know it on Earth. It is also possible that whenever life arises in the Universe some of the same forms always evolve (possibly life on a hard planetary surface always leads to traffic).

It would be very hard to detect a planet like this by taking a picture. The planet would be about one billion times fainter than Alpha Centauri A, which makes this the astronomical equivalent of trying to see a grain of sand in the direction of a bright car headlight. Detecting a planet like Jupiter would be a bit easier – Jupiter is five times further from the Sun than the Earth and much larger – but even if there were a planet like Jupiter in orbit around Alpha Centauri A, it would still be impossible to detect with current instruments. The first exoplanet was discovered using a much sneakier method.

The method works because stars wobble. As every child learns at school, the Sun is at rest in the center of the Solar System and the planets are held in their courses by the gravitational force between them and the Sun. But this is one of those many times when the teachers lied. If the gravitational force of the Sun pulls on a planet, the gravitational force of the planet, like two players in a tug-of-war, also pulls on the Sun – and so the Sun is not quite at rest. The Sun and the planets actually all orbit the center-of-mass of the Solar System. In the Alpha Centauri system, the center-of-mass is in empty space, but in the Solar System the center-of-mass is only just above the surface of the Sun (Figure 3.2). The teachers therefore only told a white lie, because for most purposes it is acceptable to say that the planets are orbiting a stationary Sun; but to be precise (and science is often a matter of details), the Sun is actually moving in a small orbit of its own. One way then to search for planets around a star would be to take many pictures of the star and look for small changes in the star's position caused by the gravitational effect of the planets. But unfortunately these changes are also too small to be detected with current technology. Astronomers finally succeeded in detecting exoplanets by using one of the oldest and most powerful techniques in astronomy.

One of the minor tribulations of being a scientist is that sometimes, inevitably, you make mistakes. For those of us who in the course of our careers have made the occasional embarrassing mistake, one consolation is that at least we have never made one as bad as the French philosopher Auguste Comte. At the beginning of the nineteenth century, Comte argued that the distances between the stars were so incomprehensibly large that we would never know what the stars are made of. His timing was superb because, almost as he wrote, scientists in German laboratories were developing the technique which would not only tell us the

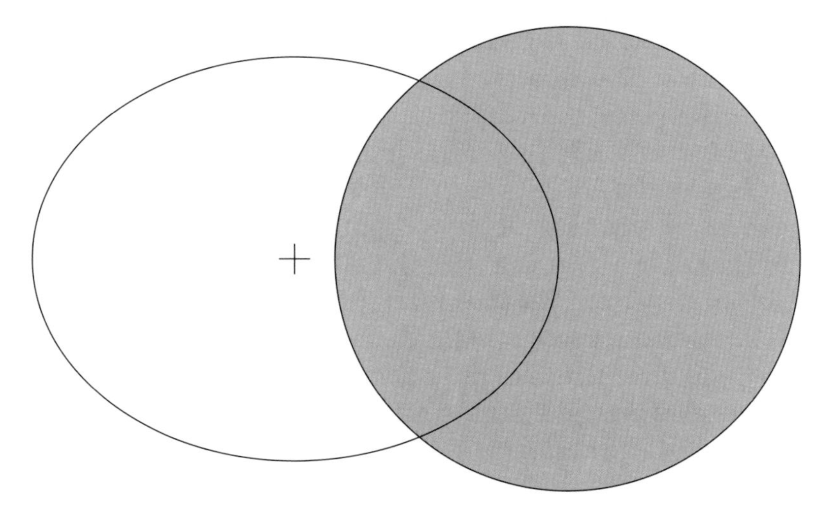

FIGURE 3.2 The orbit of the Sun. The cross marks the center-of-mass of the Solar System around which the Sun travels.

composition of the stars, but also their temperatures and how fast they are moving.

A poetic place to start is the pot of gold at the foot of a rainbow. The gold is fairy gold and, as you head towards the foot of the rainbow, the pot of gold never gets any closer. The prosaic reason for this is the laws of physics governing rainbows. Rainbows occur because sunlight contains light with a mixture of wavelengths; droplets of water in the atmosphere bend the light of different wavelengths by different amounts; and the human eye perceives the light of different wavelengths as different colors. For the rainbow to be seen, the droplets and the Sun and the human eye have to be at just the right angles. If you head towards the pot of gold, the angles are no longer correct and the rainbow vanishes. You still see a rainbow because a new set of droplets is at the correct angles, but the new rainbow – and the pot of gold – is just as far away as before.

Raindrops are a natural way of dividing light into its constituent wavelengths, but you can do the same thing with a triangular piece of glass called a prism. When the German master optician Joseph Fraunhofer used a prism to analyze sunlight in 1814, he discovered that superimposed on the regular rainbow of colors were thin dark lines. These lines remained an enigma for almost fifty years. Then other German scientists discovered that each chemical element absorbs light at a characteristic set of wavelengths. They realized that the dark lines in the spectrum of the Sun are produced by the different chemical elements in the Sun. By measuring the wavelengths of the spectral lines it is possible to tell what the Sun is made of. The Sun contains every element, including gold. And so the gold is not at the end of the rainbow – it is in the rainbow itself.

This is the astronomical technique of spectroscopy. By looking for the characteristic sets of dark lines in the spectrum of a star – effectively the fingerprints of the chemical elements – it is possible to discover the composition of the star, the thing Comte claimed it would never be possible to do.

It is also possible to use spectroscopy to measure the speed of the star. To see how this works, imagine you are standing at the side of the road and an ambulance is racing towards you. If you listen carefully you will hear the pitch of the siren change as the ambulance passes – it will suddenly get lower. The reason for this is that when the ambulance is coming towards you, it is travelling in the same direction as the sound wave, effectively chasing it, with the result that the wave is compressed; the distance between two wave crests (the wavelength) is reduced and the frequency or pitch is increased. Once the ambulance has passed, it is moving in the opposite direction to the sound wave, with the result that the wave is stretched out; the wavelength is increased and the pitch reduced. This is an example of the *Doppler effect*, named after the nineteenth-century scientist Christian Doppler. This effect also applies to other kinds of waves such as light. If a star is moving away from the Earth, the wavelengths of all its spectral lines increase; if the star is moving towards the Earth, the wavelengths of all the spectral lines decrease. By measuring the change in the wavelengths, it is possible to discover whether the star is moving away from you or towards you and also its speed (and you can still find out the composition of the star, because although the wavelengths of the lines have changed, the characteristic pattern of lines for each element – its fingerprint – stays the same).

Spectroscopy is therefore another method of looking for the gravitational effect of unseen planets. Imagine that you are on the right-hand side of Figure 3.2 observing the Sun from a great distance. As the Sun orbits around the center-of-mass of the Solar System, it will move alternately towards you and away from you. Because of the Doppler effect, the spectral lines in the Sun's spectrum will move backwards and forwards in wavelength. By observing this oscillation in wavelength, an astronomer on the planet around Alpha Centauri A would be able to tell that the Sun has planets.

Human astronomers used this method successfully for the first time in 1995. The star *51 Peg* is in the constellation Pegasus and is just bright enough to be seen with the naked eye. It is about fifty light years away and its mass and color are almost exactly the same as the Sun. In that year, two astronomers at the Geneva Observatory in Switzerland, Michel Mayor and Didier Queloz, discovered that the spectral lines in 51 Peg are oscillating in just the way expected if the star has a planet (Figure 3.3). The lines are oscillating rather quickly, backwards and forwards in wavelength every four days, and the astronomers calculated that the planet must be very close to the star, only about one twentieth the dis-

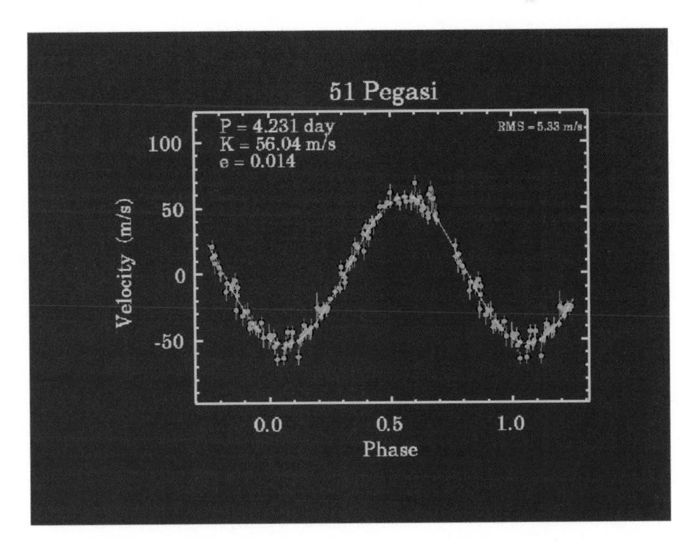

FIGURE 3.3 The variation in the velocity of the star 51 Peg caused by an unseen planet. Credit: Michel Mayor

tance that the Earth is from the Sun. They were also able to estimate the mass of the planet: about half the mass of Jupiter.

This was unexpected. In the Solar System, the giant planets are much further from the Sun than the Earth. The discovery of a giant planet so close to a star showed that the anthropocentric fallacy also applies to planetary systems: our Solar System should not be regarded as the pattern for all planetary systems. Since the discovery of the planet around 51 Peg, 155 other exoplanets have been discovered, many of them by a team led by the American astronomers Geoff Marcy and Paul Butler, and it is certain that by the time this book is published many more will have been discovered*. They have all been discovered by the same method and they are all giant planets, with masses between one fifth and eleven times the mass of Jupiter. Almost all of them are much closer to their star than Jupiter is to the Sun. At first sight, this suggests that the Solar System is actually an extremely unusual planetary system. This made some astronomers wonder whether there might be another explanation for the oscillating spectral lines. Perhaps the stars themselves are expanding and contracting, which would also make the spectral lines oscillate. Five years after Mayor's and Queloz's discovery, however, another discovery showed that the oscillating lines are definitely caused by the presence of planets.

There is one situation in which the Doppler method does not work. If the plane of the planetary system is at perfect right angles to the line joining the star to the Earth (Figure 3.4), the star's motion is never in our direction, and so the wavelengths of the spectral lines will not change. For most stars we have no idea of the orientation of the planetary system, which leads to some uncertainty in the estimate of the mass of the planet*. In one special case, however, there is another way of telling that the star has a planet. Suppose the orientation of the planetary system is

---

* Up-to-the-minute information on exoplanets, as well as on the other subjects of this book, can be found on the book's website (www.originquestions.com).

* Because of the uncertainty about the orientation, the mass estimate produced by the Doppler technique is actually a lower limit on the mass, and so the true value of the mass may actually be higher than the Doppler estimate.

FIGURE 3.4 A planetary system for which the Doppler method does not work. The planetary system is being viewed edge-on. The large circle is the star; the small circle the planet. The planet moves up and down along the line, as does the star but over a much smaller range. The spectral lines of the star do not show a Doppler effect because the star is moving at right angles to the direction of the Earth.

flipped by ninety degrees from that shown in Figure 3.4 so that the planet passes between the star and the Earth. During some of its orbit, the planet will mask part of the star. It will not block very much of the starlight because planets are much smaller than stars, but it may block enough for the brightness of the star to change noticeably; as the planet moves across the star's disk the brightness of the star will dim and then return to its normal value when the planet is no longer obscuring the disk (Figure 3.5).

Five years after the first detection of an exoplanet, an international team of astronomers reported the results of a monitoring program on the star HD 209458. Earlier observations using the Doppler method had shown that HD 209458 has at least one planet. The team found that the star's brightness varies in precisely the way expected if a planet is passing between us and the star (Figure 3.6). This discovery showed that the oscillating spectral lines are not caused by pulsations of the star or by some other strange stellar phenomenon: the star really *does* have a planet. (The team had found no dip in the brightness of other stars with Doppler detections, but this was only to be expected because only roughly one planet in a hundred will pass directly between its star and the Earth.)

Because the team had the results of both planetary-detection methods, they were able to investigate the properties of the planet in more detail than was possible for the other exoplanets. The

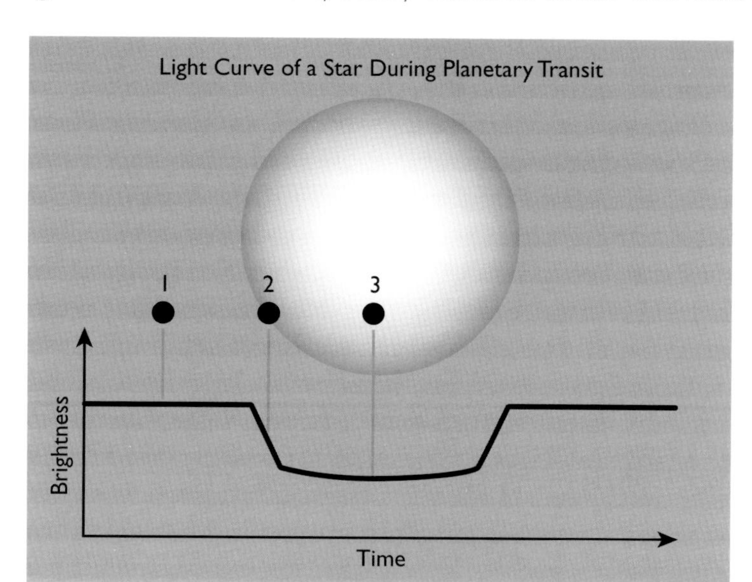

FIGURE 3.5 A prediction of how the brightness of a star would change if a planet passed between the star and the Earth.

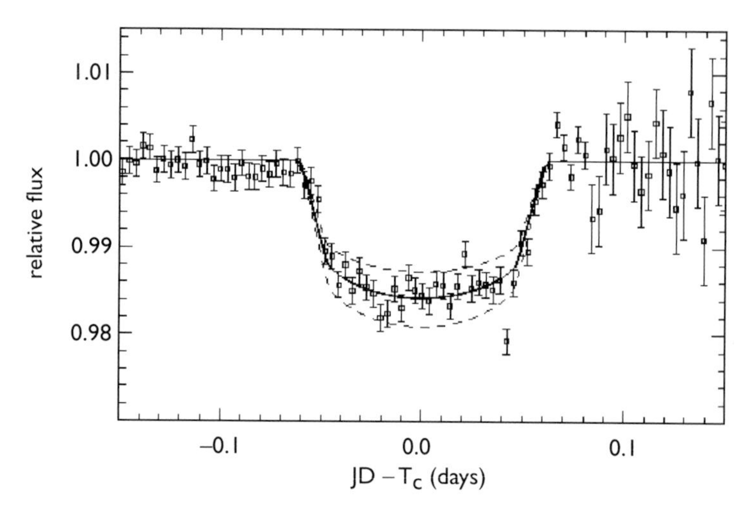

FIGURE 3.6 Variation in the brightness of the star HD 209458. The brightness drops in exactly the way expected if a planet has moved between us and the star, returning to its usual value when the planet is no longer obscuring the star's disk. Credit: David Charbonneau

planet is an extreme example of a "hot Jupiter," a giant planet that is very close to its star. The distance between the planet and HD 209458 is actually only one eighth of the distance between Mercury and the Sun. The planet must be very hot, and its temperature is probably high enough to boil iron. It has a larger diameter than Jupiter but a lower density and mass (its density is actually lower than water, which means that like Saturn in our own planetary system the planet would float if only we had an ocean large enough).

Although almost 200 exoplanets have been discovered, the results of the planetary detection experiments actually imply that most stars do *not* have planets. Of the stars similar to the Sun which have been investigated with the Doppler method, only five per cent have planets. In particular, many of the nearby stars, the natural places to emigrate if we ever develop interstellar space flight, appear not to have planets. Tau Ceti doesn't have a planet, Alpha Centauri doesn't have a planet – of the nearby stars, only Epsilon Eridani appears to have a planet, and even this is uncertain because the star has some unusual magnetic properties which make the Doppler method untrustworthy.

However, suppose that you are an astronomer on the planet around Alpha Centauri A who is looking for planets around other stars. You are particularly interested in the yellow star four light years away, because it is quite similar to your own and so may have planets suitable for hosting Alpha-Centaurian-like life (although, being a scientist and not the kind of person who reads the supermarket tabloids, you do realize that life elsewhere may be radically different from life on your own planet – its metabolism might be based on hydrogen fluoride rather than hydrogen chloride and it might even have eleven arms rather than thirteen). Full of anticipation, you observe this star using the Doppler technique. However, after a month of careful observations, you have seen no change in the positions of the star's spectral lines. You conclude that disappointingly the star does not have any planets.

The reason for this surprising result (the Solar System does, after all, contain several large planets) is that the size of a star's wobble does not only depend on the mass of the planet. It also depends on how far the planet is from the star: the further the planet is from the star, the weaker the gravitational force between the two, and the smaller the change in the wavelengths of the star's spectral lines. The giant planets close to stars – the hot Jupiters – were discovered fairly easily, because they are both big and close to their stars, and thus produce whopping Doppler signals. Planetary systems like our own are much harder to discover. With current (2006) human technology, it is possible that the Alpha Centaurian would just about be able to discover Jupiter, but he/she (it?) would fail if Jupiter were slightly further from the Sun or slightly smaller. Planets like the Earth, of course, with masses less than one hundredth the mass of Jupiter, would be impossible to detect.

Therefore, it is still possible that the planetary systems containing the hot Jupiters are the rare few per cent and most planetary systems, the silent majority, are like our own.

Nonetheless, even if only a small percentage of planetary systems have giant planets close to their stars, it is a bit surprising that there are *any*. In the first chapter I described how the heat from the newly formed Sun prevented substances with a low melting point from freezing in the central part of the solar disk. There was therefore less solid material close to the center, which naturally led to smaller planets. The outer planets in the Solar System are mostly composed of gas rather than solid material, but if they were formed by the core-accretion method (Chapter 1), they should contain rocky cores which are still much larger than the masses of the inner planets. The pattern of the Solar System – small planets close in and large planets further out – is therefore exactly what one expects if the standard model of the formation of a planetary system is correct.

One possible explanation of the hot Jupiters is that giant planets are not formed by the core-accretion method, which was anyway one of the more uncertain parts of the story, but instead by the sudden gravitational collapse of large sections of the disk. If this alternative storyline is true, there are no rocky cores at the centers of the gas giants in the Solar System, and the amount of solid material in the solar disk is irrelevant. The obvious way to test this idea would be to look for the rocky cores, but unfortunately their atmospheres are so thick that we currently have no idea whether Jupiter, Saturn and the rest have rocky cores or are gas throughout.

The more popular explanation for the existence of the hot Jupiters is that something we tend to take for granted is actually not correct: the planets are not fixed in their courses. In the last chapter I described the evidence that the objects in the Edgeworth–Kuiper Belt cannot have been formed where we see them today, beyond the orbit of Neptune, but must have been formed in the inner Solar System. Even after the formation of the planets, there would still have been billions of planetesimals, chunks of rock about 100 km in diameter, in the inner Solar System which had not been incorporated in any planet. However, for the left-over planetesimals the formation of the giant planets was notice to leave; sooner or later, a planetesimal would inevitably have strayed too close to one of the giant planets, with a high probability of being evicted from the inner Solar System by the planet's gravitational force. These gravitational encounters were probably the cause of the formation of the EK Belt and the Oort Cloud, but they may also have had a large effect on the orbits of the planets themselves.

If a planetesimal was forcibly moved further from the Sun as the result of a gravitational encounter with a giant, it would have gained gravitational energy, in exactly the same way that a man climbing a ladder gains gravitational energy as the result of moving higher in the Earth's gravitational field. The energy gained by the planetesimal must have come from somewhere (there is no such thing as a free lunch), and the only place it can have come from is the planet. Any encounter that increased the gravitational energy of a planetesimal, moving it onto a path that took it further from the Sun, must have reduced the gravitational energy of the planet, moving it onto a path that took it closer to the Sun. The change in a planet's orbit as the result of one encounter would have been minuscule, but the result of hundreds of thousands of encounters could have been a significant change in the planet's orbit. In the Solar System, because there are four giant planets, the results of these gravitational interactions will have been complex, and recent computer simulations[12] suggest that while Jupiter has moved inwards from its original position, Neptune may actually have originally been closer to the Sun. A natural explanation of the existence of hot Jupiters is if the degree of *planetary migration* varies from system to system. The reason for the variation probably lies in the amount of material originally in the disk around the star: the more material there was in the disk, the larger the number of planetesimals, and the further the planets will have moved.

From one point of view, Jupiters, whether they are hot or cold, are not very interesting. The only life of whose existence we are certain is on the surface of a small rocky planet, and so planets like this seem more interesting than gas giants. It is possible that we are being anthropocentric. But it is also possible that a planetary surface, a complex interface between solid, liquid and gas – a much more complex environment than a gas cloud – is exactly the kind of place where life is likely to start. Although we have no idea at the moment whether nearby stars are orbited by small rocky planets, we may have the answer sometime during the next decade.

The way scientists are planning to overcome the huge problem of seeing such faint objects close to bright stars (grains of sand at the edge of a car headlight) is to play a clever trick. Although professional optical astronomers normally use telescopes which do not look much different from the telescope an amateur might use in her back garden – just much bigger and usually on a high mountain – radio astronomers often combine the signals from many individual telescopes to produce a multitelescope instrument called an interferometer. The advantage of an interferometer is that it allows astronomers to see much more fine detail in the sky than is possible with a single telescope. The clever trick is to use an interferometer to turn off the headlight. By carefully tuning a two-telescope interferometer, it is possible to arrange that the signal from a star detected by one telescope is the opposite of the signal detected by the other telescope – and so when the two signals are combined they will cancel out. The signals from any object at a small distance from the star will not be cancelled out, and so the interferometer will produce an image of the planetary system, but without the star in the center.

Both NASA and ESA are planning space missions to look for Earth-like planets. The ESA mission, Darwin, will be an interferometer and is scheduled to be launched in 2015. Darwin will look for planets in infrared light rather than in optical light, because the difference between the brightness of a planet and star is

FIGURE 3.7 Artist's impression of what Darwin will look like. Credit: ESA

smaller at infrared wavelengths. The reason why Darwin will be sent into space is that the Earth's atmosphere is largely opaque to infrared light, and moreover the Earth itself and everything on it is a bright source of infrared radiation. ESA's current plan is that Darwin will consist of between four and six relatively small telescopes, each with a mirror one meter in diameter, separated by about fifty meters (Figure 3.7). Since the Earth itself is such a beacon of infrared radiation, Darwin cannot simply be placed in orbit around the Earth like the Hubble Space Telescope; the plan is that Darwin will be placed in the outer Solar System, somewhere around the orbit of Jupiter. Darwin should have the sensitivity to detect even planets as small as the Earth. Figure 3.8 shows a simulation of what Darwin would see if it were pointed at a planetary system exactly like the Solar System thirty light years away. There is a gap in the middle where the star would normally be; Mercury is a bit too faint to detect; but Venus, the Earth and Mars can all be seen.

Of course, merely detecting a planet does not immediately tell us whether there is any life on it. Nevertheless, once you

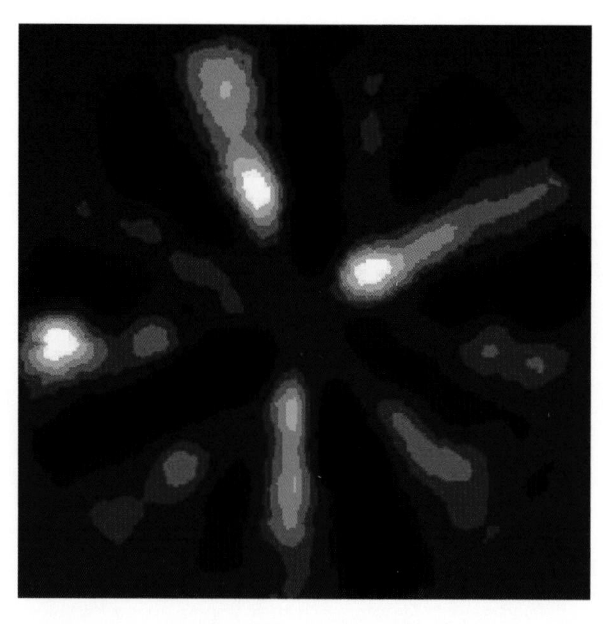

FIGURE 3.8 Simulation of what Darwin would see if it were pointed at a planetary system like the Solar System at a distance of 30 light years. The three blobs of light are "Venus," the "Earth" and "Mars" (Mercury is too faint to see). Credit: Bertrand Mennesson

can *see* a planet, there is a surprisingly easy way to answer this question.

The clue to this method is the orange-red film that can usually be seen on any iron object that has been exposed to the weather – a spade, a bicycle, a padlock on a garden shed. Oxygen is one of the most reactive chemical elements, and it inexorably reacts with the iron artefacts of our industrial civilization, forming iron oxide or rust. Oxygen is so reactive that the first production of large amounts of oxygen by photosynthesis billions of years ago posed a huge problem for life on Earth, and evolution had to generate many new metabolic systems to avoid life being destroyed by the new poisonous gas in the atmosphere. Because the gas is so reactive, in a planet's natural state there should be virtually no free oxygen. If for some reason all life on Earth was suddenly wiped out, all the oxygen would gradually vanish from the atmosphere, combining with the other atmospheric gases, rocks on the ground and abandoned bicycles.

Darwin will be able to tell whether there is oxygen in the atmosphere of a planet by looking for the characteristic spectral lines of a variant of oxygen, ozone. Once we can see a planet, if we detect these spectral lines, we will know that the planet is not in its natural state. Therefore, in about a decade, we may know not only whether there are planets like ours around nearby stars but also whether they contain life.

Nevertheless, even if Darwin is successful and finds a planet with an oxygen-rich atmosphere, we will still not know the answer to the big question. If we discover a planet with an oxygen-rich atmosphere and send out a space mission to meet our new cosmic neighbors, it would be a bit of an anticlimax to discover that the only thing living there are algae. Project Ozma was the first attempt to answer the really big question: are we alone in the Universe? Since that time scientists have continued to try to contact extraterrestrials (in addition, of course, to the regular encounters reported in the *Weekly World News*).

NASA tried to do this in the early 1970s when it sent the Pioneer 10 and 11 spacecraft to visit the outer planets. Frank Drake and Carl Sagan, the famous astronomer and science writer, realized that these would be the first human artefacts to leave the Solar System. They decided that each spacecraft should contain a message in case any extraterrestrial happened to find it. They designed a plaque for each spacecraft showing what we look like and also our address: the Orion spiral arm; the third rock from the Sun. The later Voyager spacecraft also contained a record on which, among other things, there is the sound of a baby crying. The chance of an extraterrestrial ever finding one of these spacecraft, a bottle cast upon the cosmic ocean, is minuscule; it will take a hundred thousand years to cover the distance to even the closest star, and none of the spacecraft is actually heading towards a star.

The chance of Project Ozma being successful was even smaller than the chance that one of the Pioneer or Voyager bottles will ever be picked up. The chance was always going to be slim because Frank Drake had to choose one frequency out of a billion possible frequencies and two stars out of the three hundred billion stars in the Galaxy. However, what if everyone is listening and nobody is talking? Astronomers have discovered, using the

Doppler method, that one of Drake's two targets, Epsilon Eridani, possibly has planets. Suppose that for this planetary system the dice have rolled the right way: life got started; it did not stop at the unicellular stage; and an advanced civilization now exists on one of the planets. Even with this very improbable roll of the dice, Frank Drake would still have had to have been very lucky. The civilization would have to have chosen, during the two month period of Project Ozma, not only to *transmit* a message but also to send it to the Sun out of the three hundred billion stars in the Galaxy. This seems about as likely as a castaway on a desert island walking down to the beach one morning and finding a message in a bottle addressed to himself.

However, until we invent a practical method of interstellar spaceflight, or unless the *Weekly Word News* is right and extraterrestrials are already here, radio searches are the only game in town. In the four decades since Project Ozma, there have been many radio searches for extraterrestrials – all have been unsuccessful. Advances in radio technology mean that it is now possible to listen simultaneously on a billion frequency channels. But what if everyone is listening and nobody is talking? Even this problem is on the verge of being overcome. Ironically, governments have chosen just this moment to decide that SETI projects (**S**earches for **E**xtra**T**errestrial **I**ntelligence) are no longer worth the money.

It has to be admitted that SETI projects are not entirely respectable. Part of the problem is the radically different values of SETI scientists and of mainstream scientists. A conventional scientist tackles problems that he thinks he can solve; solved problems mean research grants, academic tenure and promotion. A scientist starting a career in SETI knows that he may well spend his whole career without discovering any extraterrestrial life, but thinks that the voyage of exploration is enough even if there is no landfall at the end. SETI scientists are dreamers. The other part of the problem is the *Weekly World News*, *Star Trek* and science fiction movies. Every time a committee of worthy scientists and politicians meets to consider a SETI project, it is impossible for them not to think of ET riding a bicycle in the sky over the Californian hills and headlines like "Aliens stole my mother." As Congressman Silvio Conte of Massachusetts said when NASA, the main patron of SETI research, was seeking budgetary approval of its SETI program in 1990: "Of course there are flying saucers and advanced civilizations in outer space. But we don't need to spend $6 million this year to find evidence of these rascally creatures. We only need 75 cents to buy a tabloid at the local supermarket." Three years later, Congress finally cancelled the U.S. government funding of the SETI program.

Fortunately, as everyone knows, California is full of eccentrics – and some of those eccentrics are billionaires. The SETI program is continuing under the auspices of the SETI Institute, a privately funded organization in Mountain View, California. Most of the money for the institute comes from wealthy individuals in the computer industry who, depending on your point of view, are either mildly deranged or have enough imagination to understand the importance of SETI and enough money not to care that they may see no tangible results in their lifetime.

A major goal of the SETI Institute is to overcome the problem experienced by previous radio SETI projects of having to cadge time on regular radio-telescopes, which is often difficult because of the competition from respectable astronomy projects. The Institute plans to build a radio-telescope consisting of 350 satellite dishes (satellite dishes are cheap) in the desert in California. The whole telescope will cost about 25 million dollars, which will be paid by Paul Allen, the cofounder of Microsoft. The Allen Telescope Array will be used only for SETI projects. It will also be so sensitive that it may overcome the "Is everyone listening and nobody talking?" problem.

We may not be talking with ET at the moment, but we are babbling to the Universe all the time. Artificial radio signals have been leaking into space since the beginning of the radio age. These signals are much fainter than the signal we could send if we aimed a signal at a particular star, but they are being transmitted in all directions. If there is a civilization around Alpha Centauri which is similar to our own, it too will be inadvertently broadcasting radio signals. With a large enough radio-telescope, it should be possible to eavesdrop on these signals – whether the Alpha Centaurians are interested in talking to us or not. If we are lucky, if these signals are being broadcast by a civilization around a nearby star at a slightly higher level than currently from the Earth, the Allen Telescope Array will be able to detect them.

The message we are broadcasting is a much less elegant one than the one devised by Frank Drake and Carl Sagan. The message in the bottle on the Voyager spacecraft included such cultural treasures as a recording of all Bach's Brandenburg Concertos. The message we are broadcasting is our daily diet of television and radio (the astronomer with thirteen arms on the planet around Alpha Centauri will need a TV guide from the year 2002). But if it is less elegant, possibly an alien would learn more about what human life is really like from the TV soaps than from the Brandenburg Concertos. I wonder whether it is a coincidence that UFOs started being reported a few years after the first television broadcasts.

Part II

Stars

4. Connections

It is difficult to see the change in a planet during a human lifetime, but it is not impossible. Anyone who lives near a volcano has seen a planet's surface being reshaped. Anyone who lives near a river is a bystander, even if they don't realize it, to the slow-motion destruction of a mountain. Boil the water taken from any river and you will find left at the bottom of the pan a tiny residue of dirt, dirt that has been washed off the hills and mountains by rain. Over millions of years, this slow erosion can reduce the height of a mountain by thousands of meters (the Grampian Mountains in Scotland were once the size of the Himalayas). The story of a planet is a slower-paced story than our own human stories, but it is possible, if one looks carefully, to see it happening in front of our eyes. It is much harder though to see the change in the stars. The night sky looks the same from generation to generation and it really does seem to be the eternal backdrop to our individual stories. The stars we see today are the same as the stars that the Romans and Greeks saw, and are even virtually the same stars our ancestors saw in Africa one million years ago*.

One day though in the sixteenth century somebody did see a change in the night sky. In 1572, a young Danish nobleman, Tycho Brahe, was walking through the woods one evening around Herrevad Abbey, where he was staying with his uncle Steen Brahe. Tycho and his uncle were members of a family that was one of the pillars of the Danish monarchy, but Tycho was not a typical aristocrat, concerned only with land, power, fighting and hunting.

* The positions of the stars are gradually changing. Our ancestors one million years ago saw the same stars, but they did not see the same constellations we see today.

Like many young aristocrats in all countries at this time, Tycho was imbued with the ideals of the *Renaissance*, the rebirth of learning that had started a century before in Italy. He already knew at least four languages and had just returned from eight years of studies abroad. He was now staying with his uncle at the abbey, where they had started several technical and scientific projects: a paper mill, a glassworks, an instruments factory, a chemical laboratory and an astronomical observatory.

While walking in the woods that evening, Tycho noticed something different about the constellation Cassiopeia, the irregular W of stars not far from the Pole Star. There was a bright new star not far from the upper right star in the W. The appearance of this new star changed Tycho's life, and it is not an exaggeration to say that it also shattered the medieval view of the Universe.

The author of this view, although he had lived two millennia earlier, was the Greek, Aristotle. The Universe, according to Aristotle, consists of a complicated set of spheres centered on the Earth. The Earth is at rest, which is only common sense because otherwise we would all fall off. The daily cycle of night and day is caused, in this view, by the rotation of the spheres around the Earth rather than the rotation of the Earth on its axis. The spheres, which were often assumed to be made of transparent crystal, transport the objects we see in the sky around the Earth. The lowest set of spheres carries the Moon, the next the Sun, and the following five the planets: Mercury, Venus, Mars, Jupiter and Saturn. The outermost sphere is the one on which the stars are fixed. Aristotle claimed that outside this sphere must sit the *Prime Mover* who sets the spheres in motion. An important part of Aristotle's teaching was the distinction between the Earth and the heavens. Objects on Earth tend to move up or down; objects in the heavens move in perfect circles. Objects on the Earth are made of a mixture of four elements – earth, air, fire and water; objects in the heavens are made of an incorruptible fifth element or essence (which later gave us the word *quintessence,* from the Latin for five, *quinque*). This idea of a celestial region – incorruptible, unchanging, and separate from the world of change and decay we see around us – naturally had a strong appeal to the medieval Church. Aristotle's ideas were not the only ones circulating in the sixteenth century. A tiny minority would have heard of the revolutionary idea of

Copernicus that the Sun rather than the Earth was at the center of the Universe. However, it was the ideas of Aristotle that formed, if they ever thought about such things, the cosmological backdrop to most people's lives; and it was also these ideas which had received the imprimatur of the Church, which made questioning them sometimes a hazardous business.

The appearance of this new star was a challenge to the supremacy of Aristotle's views, because if he was correct the night sky should be eternal and unchanging. However, according to Aristotle, anything on the Earth or below the spheres of the Moon is subject to change. Therefore, it was possible to evade the challenge by assuming that the new star was not like the other stars, but was instead a phenomenon in the Earth's atmosphere or somewhere else in the sublunary (below the Moon) region. In previous generations, this is almost certainly the assumption that would have tacitly been made. Evidence for this is the medieval reaction to another transitory phenomenon in the night sky – the appearance of a comet; it was always assumed that the comet was in the Earth's atmosphere and therefore part of the sublunary region of change and decay. Now however, as the result of the Renaissance, there was a new spirit abroad in Europe.

The zeitgeist of the sixteenth century, the spirit that united the achievements of Tycho, Galileo, Leonardo and virtually any scientist or artist one can think of, was experiment*. The Greeks, in contrast, had not been great experimenters and their achievements had been made by a mixture of observations and theory. To realize the difference in the spirit of the times, one has only to read one of the dialogues of the Greek philosopher Plato and compare his appealing description of a world in which deep intellectual debates would suddenly start in the street, but in which nobody ever mentions testing anything, to the practical scientific and technical activities underway at Herrevad Abbey. When Tycho saw the new star in Cassiopeia in 1572 he immediately thought of a way to test the idea that it was in the sublunary region. Within a few years he had also applied the method to a

* At first thought, an obvious exception is Copernicus, but he actually spent many years making observations to test his theory.

comet. He showed that both the new star and the comet are not in this region, but are instead in the celestial region that, according to Aristotle and the Church, should be incorruptible and unchanging.

The idea behind this method is simple. As I write this chapter, I am sitting at a table in a room at the back of my house overlooking the garden. It is a pleasant sunny May afternoon and the hawthorn tree at the bottom of the garden is in flower. In the distance, there is a line of trees marking the edge of a park, which provides a backdrop to my view out the window. From where I am sitting, the hawthorn tree appears to fall between two of these distant trees. If I get out of my chair and move to the right side of the window, the hawthorn tree appears to have moved; it now lies in front of one of the distant trees. This apparent shift in the position of nearby objects relative to distant objects is the phenomenon of *parallax.* The size of the shift depends on the distances of the objects. On the window there is a grid of metal strips. To see a change in the position of one of these strips relative to the distant trees, I don't even have to get out of my chair; I just have to move my head slightly.

The reason that parallax is so important in astronomy is that by measuring the shift in the position of an object relative to some very distant objects, you can estimate the distance of the closer one. This does require you to know the distance between your two viewing positions – the distance from one side of the window to the other, for example – but other than that, all you require is some basic high school trigonometry. The distances of all objects in the Universe outside our planetary system, from the closest stars to the most distant galaxies, are ultimately based on this method.

Tycho applied this method to the new star. As the Earth rotates, we are moving and so the point from which we are viewing the Universe is changing. Tycho realized that if the star was in the sublunary region, its position relative to the stars painted on the eighth sphere would change with our vantage point (Figure 4.1). At this point, the knowledgeable reader may object that at the time virtually everyone, including almost certainly Tycho, assumed that the Earth is actually at rest, and that rather than the Earth spinning on its axis, the Universe is revolving around the Earth – and if this were true, our vantage point would not be

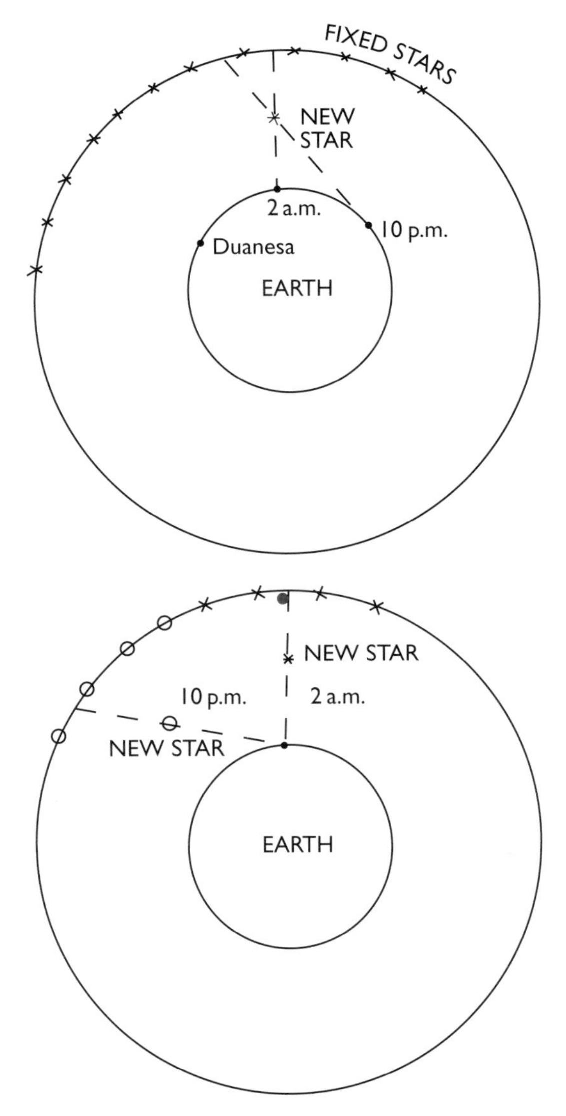

FIGURE 4.1 The expected parallax of Tycho's new star if it is below the sphere of the fixed stars. In the top diagram the Earth is rotating; in the bottom diagram the Universe is rotating around the Earth. The dashed lines show the observer's sightlines at 10 p.m. and at 2 a.m. In the top diagram the observer's position has moved, and the new star appears to lie between different pairs of distant stars at the two times. In the bottom diagram, the observer stays in the same place and all the stars have moved (the circles and the crosses show the stars' positions at the two times). The new star still appears to lie between different pairs of distant stars at the two times.

moving. However, even if the Aristotelian spheres are rotating around a stationary Earth, one expects to see shifts in the positions of objects on the inner spheres relative to those on the outer spheres (Figure 4.1). Tycho used his astronomical instruments at Herrevad Abbey to measure the position of the new star relative to other stars. He found that as the night progressed the new star did not move relative to the other stars. From the absence of movement, he argued that the new star could not be in the sublunary region. A few years later, in 1577, he carried out the same measurements for a comet, again finding no change in its position relative to the fixed stars. These two sets of measurements showed conclusively that there is not a separate incorruptible celestial realm, but that the objects in the night sky are part of the same world of change and decay we see around us.

Tycho's observations of the new star of 1572 changed his life in an important practical way. He dashed off an account of his observations, which he sent to the king of Denmark, King Frederick II. The king was already well disposed towards Tycho because of his family connections and he gave Tycho the financial resources to set up an astronomical observatory on the Island of Hven. The observatory that Tycho created on the island, Uraniborg, from which he and a team of assistants made careful measurements of the sky for the next twenty years, was the prototype of all modern observatories. Many of the procedures adopted at Uraniborg are now so much part of the fabric of modern observational astronomy that we observers tend to take them for granted: the repetition of observations to reduce errors; independent measurements by different observers as a guard against personal bias; an exhaustive search for subtle effects of the instrument or atmosphere that may produce systematic errors in one's results; the importance of inventing clever new instruments as a way of increasing knowledge.

The measurements that Tycho had made of the new star of 1572 had demolished one model of the Universe. The measurements he made at Uraniborg helped to construct a new one. One of the reasons why Copernicus' heliocentric model, in which all the planets including the Earth orbit the Sun, had not been immediately adopted by everyone was the perfectly legitimate one that it did not actually reproduce very well the observed motions of the planets across the sky. However, at the beginning of the seventeenth century, one of Tycho's former assistants, Johannes Kepler, who had acquired his accurate measurements over many years of the position of Mars, showed that if the planets move around the Sun in ellipses rather than circles, a heliocentric model can reproduce their tracks across the sky with exquisite accuracy.

Until I sat down to do the preparatory reading for this chapter, I had always regarded Tycho, for all the reasons above, as setting the pattern for all modern observers (with just a few interesting sixteenth century embellishments – he had for example several replacement noses, the consequence of a duelling injury during his student days). When I read the account of the new star that Tycho wrote for the king, I realized I had been a little naïve.

In some places in the book, Tycho does sound like a modern scientist, making sensible speculations, firmly grounded on commonsense and observation, about the cause of the new star. He argues that space is unlikely to be empty, although the stuff that fills it must be thin enough not to impede the progress of the planets. He suggests that every so often this material may become "compacted and condensated into one Globe, and being illustrated by the light of the Sunne, might give forme and fashion to this Starre." The first part of this quotation sounds remarkably like the modern theory of how a star is born, but the rest of the quotation is enough to dispel the idea that Tycho realized the stars and the Sun are the same kind of object. He answers the question of why stars are not seen forming all over the sky by noticing that the new star is close to the Milky Way, where, he suggests, the material filling space may be unusually dense (another eerie echo of modern ideas about the formation of stars – Chapter 5).

But much more of the book is devoted to astrological speculations about the new star's significance. Tycho even discusses biblical prophecies that may refer to the new star, and he quotes a Sybelline prophecy found in 1520 in Switzerland "engraven in a Marble Stone in old latine characters" which he thinks may explain its appearance. To a modern astronomer, this book, written by the father of observational astronomy, is all rather disturbing. (A simple statistical comparison shows how inculcated the sixteenth-century world was with the ideas of astrology. Early English Books Online is a computer database of all books published in

English before 1700; searching this database, I found 232 books on the subject of astronomy and 2161 on the subject of astrology – roughly a factor of ten to one.) Nevertheless, although I certainly found Tycho's book rather perplexing on first reading, if one tries to understand his world from the inside and understand the book in the context of the knowledge available at the time, it makes a lot more sense.

For a start, Tycho would also have regarded many astrologers as charlatans. As an intelligent man, he fully realized that most astrological predictions turn out to be wrong. In the book that he wrote for the king, he argued that it is extremely hard to discern the significance of events in the sky, which is why most astrological predictions are wrong, but that it does not follow that these events have *no* significance. *We* might argue, why would the motions of large balls of gas and rock millions of miles away have any connection to human affairs? But the real natures of the planets and stars were not known in the sixteenth century. Tycho himself was strongly influenced by the ideas of the fifteenth century Italian humanists, Marsilio Ficino and Giovanni Pico della Mirandola. According to these scholars, who had adapted ideas whose original source was Plato, there is a hierarchy of three worlds with God as their source. The world closest to God is the angelic world or the world of the spirit; next is the celestial world we see in the night sky; finally there is the sublunary world. In their view, there was no sharp division between the pure celestial world and the corruptible sublunary world, and Ficino argued that the individual human is actually a fourth world containing a mixture of elements from the other three. These ideas contain two arguments for astrology. First, the celestial sphere is closer to God and so the motions of the planets and stars might be expected to reflect and announce God's plans for the Universe. Second, if Ficino was correct, there is a fundamental cosmic connection between human beings and the celestial world, because every human contains elements from this celestial world. Ficino's view was that human beings still have the power of free will, but their lives are inevitably influenced by the appearance of the sky at the moment of their birth.

If the motions of the stars and planets do have some significance, how can one discern what the stars foretell? Tycho had a robustly empirical answer. He argued that one can only determine the significance of some astronomical event by looking at what happened in human affairs the last time such an event was seen. According to the Roman writer Pliny, the Greek astronomer Hipparchus had also seen a new star. Because this previous new star was seen at the time of the rapid expansion of the Roman Republic, Tycho argued that the new star of 1572 probably foretold some similarly important change in human history. He also made the commonsense argument that since the star could be seen from many countries, it was likely that this change would affect a large part of Europe. After this he was on shakier ground. He made some conjectures based on the color of the star and its position, but he admitted that these were conjectures. His arguments about the significance of the new star based on passages in the bible also seem less bizarre when one recalls that this was the century during which scripture replaced the Church in many countries (Denmark was a Protestant country) as the primary authority in peoples' lives.

The new star appeared in the sky over northern Europe for two years and then faded from view. For the next four hundred years Tycho's star disappeared from history. Rather improbably, a second new star appeared in the sky within a generation; it appeared in 1604, when it was studied by Tycho's former assistant Johannes Kepler, which was spectacularly poor timing because this was only five years before the invention of the telescope (unfortunately, no similar object has been seen since in our Galaxy).

The true nature of stars only became completely clear three and a half centuries after the appearance of Tycho's star. Their nature was revealed by another careful observer of the sky, the German astronomer Friedrich Wilhelm Bessel. Bessel used the same method Tycho had tried on the new star of 1572: *parallax*.

Suppose that the stars are not merely points of light painted on a celestial sphere but are objects distributed throughout space. Some of these objects will be closer to us and some will be further away. As one changes ones viewing point, the closer objects should appear to move in the sky relative to the backdrop of the more distant objects. Astronomers searched for these movements for over three hundred years, but without success.

Bessel was the one who succeeded. He had two major advantages over Tycho. First, he was aware that our viewing point changes by a much greater distance than Tycho had realized. Tycho was aware that during a 24-hour period our viewing point changes because the Earth spins on its axis – or, as Tycho would have put it, because the celestial spheres rotate around the Earth. Bessel and all other astronomers since the mid-seventeenth century, when the heliocentric Universe was finally accepted, have been aware that our vantage point changes by much more than this: by three hundred million kilometers every six months because of the Earth's annual motion around the Sun. Even with such a huge change in vantage point, the movements of the stars are still minuscule. With the instruments at Uraniborg, the most sophisticated astronomical observatory of the time, Tycho had been able to measure the positions of the stars with an accuracy of one minute of arc. This is a small angle, roughly one thirtieth of the diameter of the Moon, but it is still much greater than the movements of the stars. When in 1838 Bessel finally detected the annual parallactic shift of the star 61 Cygni (Bessel's second great advantage over Tycho was the tremendous improvement in astronomical instruments during the intervening three centuries), he discovered that it is only 0.3 seconds of arc. To appreciate how small this is, imagine viewing a dime from a distance of about seven kilometres – this is how big this angle appears on the sky. With some simple trigonometry, Bessel was able to use this measurement to estimate the distance of the star: 10.4 light years. Bessel's result showed conclusively that the stars are not painted on a celestial sphere; they are huge balls of gas like the Sun (as indeed many astronomers had suspected for centuries); they appear as points of light because they are so far away.

The light year, the distance light travels in one year, feels to me quite a homely unit I suppose, because the distances of the closest stars measured in light years are less than the number of fingers on my hand. But if I translate light years into regular units, the distances become mind-bogglingly immense. In everyday units, the distance of 61 Cgyni is 98,392,320,000,000 kilometers. The only way to grasp the significance of such a large number is to construct a scale model of the Universe.

Suppose that I place a marble on the table on which I am writing to represent the Earth. Then on the same scale, the Moon would be about the size of a pea and at a distance of 30 centimeters from the marble, still just about on the table. In this scale model, Mars would be another marble, slightly smaller than the first, and it would have to be placed about 100 meters away, on the other side of the park at the bottom of my garden (a scale model would clearly be quite difficult to build in practice). The Moon is the only astronomical object that humans have ever visited and it took the Apollo astronauts five days to get there; the huge difference between the distances of the Moon and Mars in the model immediately shows why it has been so difficult making the leap to visiting the planets. In this scale model, the Sun would be the size of a beach ball and roughly the same distance from us as Mars, although in a different direction. If we discount Pluto (Chapter 2), the outermost planet is Neptune. Neptune would be about the size of a small orange and would have to be placed about four kilometers away. Even if impossible to build in practice, this scale model is a nice illustration of the sheer emptiness of the Solar System – about the size of a small city but completely empty apart from a few oranges and marbles. The Universe outside the Solar System is even emptier. In this same scale model, the nearest star – another beach ball – would be 32,000 kilometers away, making our planetary system seem comparatively a rather crowded neighborhood.

The barely fathomable void between the tiny warm neighborhoods around stars is a chilling thought, but Victorian astronomers were faced with a more substantial problem. Where do the stars get the energy to warm these neighborhoods? This is a huge problem. Almost everything on the Earth – the trees and flowers in my garden, the food in my refrigerator downstairs, the rain that is currently falling from a dreary Welsh sky, even the cars that I can hear in the distance – ultimately depends on the power of the Sun. The Sun pumps out a huge amount of energy every second and it has been producing energy to warm our planetary neighborhood at roughly the same rate for four and a half billion years. In the nineteenth century nobody knew the source of this energy. The best idea that scientists could come up with was that the Sun is gradually shrinking.

Anything in a gravitational field has energy. If an object moves down through a gravitational field, some of this gravitational energy is converted into other kinds of energy. If I push a book off my table, gravitational energy is first turned into kinetic energy, the energy of motion, and then ultimately when the book hits the floor into a small amount of heat. The Sun, of course, has an intense gravitational field. Victorian physicists suggested that if the Sun is shrinking, effectively sinking through its own gravitational field, the conversion of gravitational energy into heat might explain the Sun's vast power output. They calculated that by this process the Sun could produce energy at its current rate for about one hundred million years. In the nineteenth century the ages of the Sun and the Solar System were not known precisely, but geologists and biologists argued that the physicists must be wrong, because the fossil record and the timescales of geological processes implied that the Earth had been in existence for much longer than this. Within the pecking-order of scientific disciplines, physicists notoriously tend to despise less-mathematical disciplines such as biology and geology, and so it was fairly easy for the nineteenth century physicists to ignore these arguments. The physicists were wrong; the geologists and biologists were right.

Another Victorian mystery was the nature of the new stars observed by Tycho and Kepler. After their discovery, it had soon been realized that these could not really be newborn stars, because if so they should not have disappeared from view. It also became clear during the course of the nineteenth century that these "new stars" were an even more powerful phenomenon than normal stars. The Andromeda Nebula is a faint patch of light that can just about be seen with the naked eye. In 1885 a new star similar to Tycho's and Kepler's new stars was seen in the direction of the nebula. At the time, it was suspected, although not yet known definitely (Chapter 6), that the Andromeda Nebula is actually a separate galaxy. If the Andromeda Nebula were a separate galaxy, the new star, or the *supernova* as this phenomenon was eventually to be named, had to be producing over one billion times more energy, albeit for a much shorter time, than a normal star. Many famous and not-so-famous scientists turned their hands to producing an explanation for these new stars: a collision of two stars (Newton); a surface conflagration (Laplace); a tidal effect caused by the close approach of another star (Klinkerfues); the collision between a dust cloud and a star (Seeliger); the collision between two streams of meteorites (Lockyer). But until the twentieth century, none of these ideas was anything more than speculation.

Neither of these mysteries was solved until the twentieth century, but astronomers in the nineteenth century did make some progress in understanding the nature of the stars. Once Victorian astronomers started to systematically measure their properties, they soon realized that their apparent similarity in the night sky conceals a huge amount of variety. Some stars are emitting ten thousand times more energy than the Sun each second and others emit one thousand times less energy. Some stars are much redder than the Sun, others much bluer. The most intriguing variation however is in the spectra. As I described in the previous chapter, a star's spectrum contains information about its chemical composition, because different chemical elements produce different sets of spectral lines (the fingerprints of the element). Some stars have spectra that are dominated by the lines of hydrogen; others have spectra that contain strong lines produced by metals such as calcium; yet others have spectra dominated by the spectral lines from molecules such as carbon monoxide and water – in all, Victorian astronomers classified stars into seven spectral categories (A, B, F, G, K, M, and O) with many subcategories.

Merely dividing the spectra into classes is taxonomy or "butterfly collecting" as astronomers, mostly wannabe physicists, are unfortunately prone to call it. The crucial next step, which takes this discovery in the physicist's mind above mere butterfly collecting, is to understand the reason for the different classes. The obvious explanation, that different stars have different chemical compositions, is not necessarily true, because other things can affect the prominence of the spectral lines. After a certain amount of blundering around, astronomers realized that if they arranged the spectral classes in the order M, K, G, F, A, B, O, what they had was effectively a sequence of temperature. The crucial clue that this is correct is that this is also a sequence of color. The color of an object is a measure of its temperature; red objects are colder than blue objects. As one moves along the sequence, one moves from cool red objects to hot blue objects, and the prominence of the spectral lines of different elements changes because of the increasing temperature*. At the beginning of the twentieth

* The jumble of letters reflects the difficult history of trying to understand the underlying reason for the different classes. The disorderly jumble has also created a memory problem for generations of astronomy students. The most popular, although politically incorrect, mnemonic is **O**h **B**e **A** **F**ine **G**irl, **K**iss **M**e.

century, two scientists used this discovery to draw the most famous diagram in astronomy.

The Danish astronomer Ejnar Hertzsprung and the American astronomer Henry Norris Russell did what seems (admittedly with hindsight) a rather obvious thing. They independently discovered that if they plotted a graph of the spectral class of a star against its luminosity an interesting pattern emerged. The obvious feature of this graph (Figure 4.2) is the band of stars across the center. The existence of this band, which is called the *main sequence*, shows that, as one might expect, hot stars are generally more luminous than cold stars. The Sun, a remarkably average star (for the existence of life mediocrity is probably a good thing), is right in the middle of this band. The more interesting stars, however, are not the ones on the main sequence but the ones in the groups at the top right and the bottom left of the graph. The stars at the top right are luminous cool stars. A cool star emits less radiation from each square meter of its surface than a star that is hot. So for a star to be both cool *and* luminous, it must have a very large surface area. These *red giants* are cooler and more luminous than the Sun

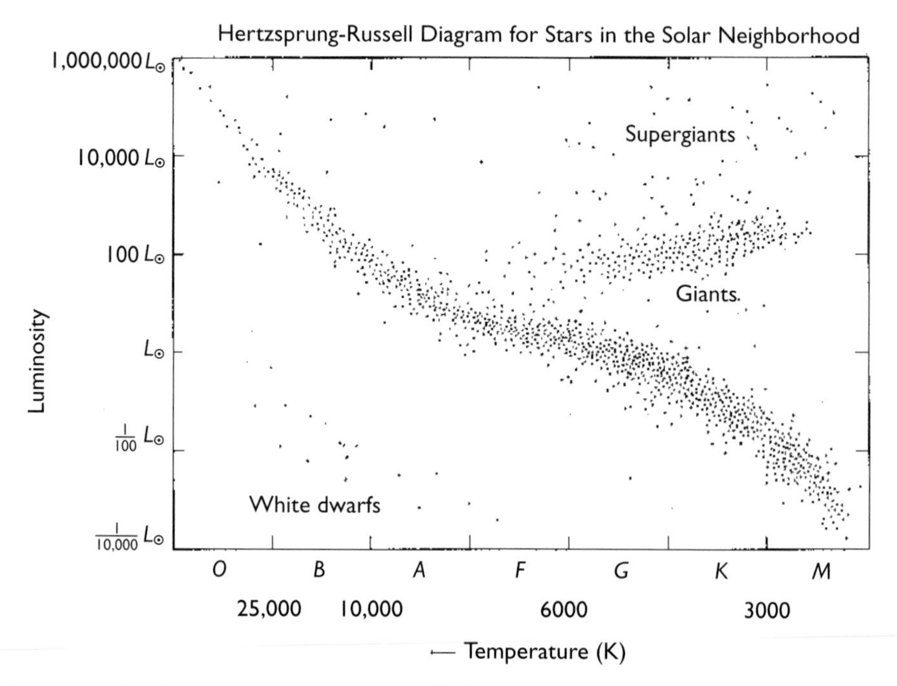

FIGURE 4.2 The Hertzsprung–Russell diagram. Credit: Gene Smith

and so they must be much bigger; if the Sun were replaced by a red giant, the Earth would be either close to the star's surface or even inside it. The stars at the bottom left are stars that are very hot but that emit, by the standard of the Sun, only a dribble of energy. With the same reasoning, these stars – *white dwarfs* – must be much smaller than the Sun.

The discovery of this pattern was an important step towards an understanding of stars, but initially nobody quite knew what to make of it. One possibility was that the pattern represents the life history of a star. Perhaps a star is born in one part of the diagram and moves from one group of stars to another during its life. An early suggestion was that a star is born at the top left of the main sequence and then, in the course of its life, gradually moves down the main sequence to the bottom right. This suggestion however did not explain the existence of red giants or white dwarfs, or indeed much else, and astronomers did not interpret this pattern correctly until the middle of the twentieth century.

In physics, a notoriously uncool subject, the only formula that is probably cool enough to appear as graffiti on the wall of a subway is $E = mc^2$. In Einstein's iconic formula, the E stands for energy, the m for mass and c is the speed of light. The first of these is a chameleon substance, but because we use it, in all its forms, so much in our daily lives – heat, light, the chemical energy in coal and gas, to name just three – energy is a concept that is not too strange. Mass is a bit trickier because of the general confusion between weight and mass: mass is the substance of an object, how much of it there is; weight is the gravitational force on the object. Bathroom scales therefore *do* measure weight, but although you could easily lose weight by taking the scales to the Moon, there would really still be just as much of you as before – your mass would be the same. Einstein showed that in certain special circumstances energy can be transformed into mass and mass into energy. His formula states how much energy is needed to create a certain amount of mass, or conversely how much mass is needed to produce a certain amount of energy. This is the solution to one of the two Victorian mysteries. The ultimate reason why the grass outside is green, why there is food in my refrigerator and why there are cars congesting the road outside is that every day some mass disappears from the Universe.

All stars are different. All stars are the same. Despite their apparent variety, all stars are at heart the same object – a much simpler object than a planet. Each star is essentially a large ball of gas held in equilibrium by two competing forces, gravity and pressure. Gravity pulls the star inwards; pressure pushes the star outwards. In the Sun, the pressure needed to balance the Sun's intense gravitational field is provided by the hot gas inside it. Any gas exerts a pressure because of the motions of the atoms or molecules in the gas. The pressure in a car tire is produced by the motions of the air molecules in the tire; each time an air molecule bounces off the tire wall it exerts a minuscule outwards force on the wall, and it is the combination of the forces produced by the millions of collisions that occur each second that *is* the pressure. The hotter the gas, the higher the pressure, and the temperature at the Sun's center is so high that the gas pressure is enough to hold up the weight of the star – in exactly the same way that the pressure of the air in the tire holds up the weight of the car.

The pressure, density and temperature at the center of the Sun are so high that a process occurs there that would have fascinated the alchemists of Tycho's time. The alchemists' ultimate aim was to learn how to transmute one element into another, especially for the obvious reason iron into gold. They failed and five hundred years of chemistry experiments have failed. It is easy enough to change one chemical compound into another, but the chemical elements that make up the compounds – oxygen, nitrogen, iron, gold, and so forth – remain stubbornly the same. At the center of the Sun, however, in a furnace hotter than any alchemist's, this long-sought transmutation occurs as a daily routine.

The Sun is mostly made of hydrogen, the simplest of the chemical elements. Each hydrogen atom consists of a proton orbited by a much smaller particle, the electron. The next most common element in the Sun is helium. A helium atom consists of a nucleus containing two protons and two neutrons, orbited by two electrons. (This image of an atom as a little Solar System may sound suspiciously simple but, as I will explain later, the subatomic world is so bizarre that this mental image is as good as any other.)

At the high temperature inside the Sun, however, the atoms do not exist as atoms; the intense heat strips the electrons away from the atoms, producing a sea of electrons, hydrogen nuclei, helium nuclei, with a seasoning of the nuclei of other chemical elements. In this high-temperature high-density sea, nuclei come much closer to each other than they do at everyday temperatures and densities. They sometimes come close enough that they interact through the strong nuclear force, a force that has such a short range that, unlike the gravitational and electromagnetic forces, it does not intrude on our notice in our everyday world. The effect of these interactions is that every so often one element is changed into another. In a chain of reactions, four hydrogen nuclei are fused to form a helium nucleus, two of the protons being converted into neutrons. A strange property of this transmutation is that the mass of the ingredients does not quite match the mass of the product. For every kilogram of hydrogen nuclei, only 993 grams of helium nuclei are produced. The missing seven grams have been transformed into energy. Seven grams does not sound very much, but in Einstein's formula mass is multiplied by the square of the speed of light, a very large number. Seven grams of matter transformed into energy is enough to supply the world's energy needs for about one hour. The Sun's energy needs are rather greater than the human race's. In the center of the Sun, 480 billion kilograms of hydrogen are transmuted into helium *every second* – of this, three billion kilograms are transformed into energy and disappear from the Universe.

The Sun is middle-aged. It has been radiating energy at approximately the current level for four and a half billion years. It contains enough hydrogen fuel to continue radiating at this level for another five billion years. After five billion years, however, all the fuel will have been used up, and the nuclear fusion reactions in the Sun's core will shut down.

If there are people living on the Earth in five billion years time, initially they will not notice much. The Sun will continue to shine; life on Earth will continue as normal. The difference will be that deep beneath its surface, the Sun, unknown to anyone on Earth, will have switched to a different source of power.

A star's backup power source is the one suggested by Victorian physicists: gravitational energy. Once all the hydrogen in the Sun's core has been transmuted into helium, the core will begin to shrink. As the core shrinks, sinking through its own gravitational field, gravitational energy will be turned into heat, which will replenish the energy the Sun is losing through radiation. This change, although unseen by anyone on Earth, is a crucial change because it will mark the point at which the Sun leaves its stable middle-age on the main sequence and enters its turbulent later years.

These years will be a sequence of surprises. The first surprise is that although the nuclear reactions in the core have shut down (there will still be nuclear reactions occurring in a thin shell of gas surrounding the core), the core's temperature will actually begin to increase. The explanation of this is that more energy will be released by the transformation of gravitational energy than is needed to replenish the energy lost by radiation; the excess energy will heat up the core. The pressure of the gas will increase as the temperature increases (the pressure in a gas, remember, is caused by the motions of the atoms, and if the gas is heated the atoms will move faster). As the core shrinks, the density of the gas will also increase. This will produce a more intense gravitational field, but the increase in pressure will stop the core collapsing.

The second surprise is that as the core slowly contracts, the outer layers of the Sun will actually expand. The explanation of this is more complicated, but an analogy that is not completely misleading is how the air in a hot-air balloon expands because of the heat from a burner suspended underneath. In this analogy, the Sun's core, which now has an extremely high temperature, is the burner, and the gas surrounding the core is the gas in the balloon; the outer layers of the Sun will expand because of the heat from the furnace at the Sun's center.

The Sun will swell until its radius is about 150 million kilometers, which by chance is the distance between the Earth and the Sun. The Sun's outer layers will cool as they expand, although they will still be hot enough to vaporize the Earth. From its current phase as a middle-aged yellow star on the main sequence, the Sun, in its first post-menopausal incarnation, will become a red giant. Although the surface will be much cooler it will be much larger, and as a consequence the Sun will be more luminous than it is today. This is the third surprise: although it has lost its nuclear energy source, the Sun will become a more luminous star.

These events are far in the Sun's future, but Betelgeuse, the top left-hand star in the constellation Orion, is already a red giant. Betelgeuse, which has a much higher mass than the Sun, is so large that if placed at the center of the Solar System, it would engulf not only the Earth but also Mars and Jupiter. As viewed from the Earth's surface, the images of all stars have angular sizes of about one arc second (the angular size of a dime viewed from a distance of two kilometers). This is caused by the blurring effect of the atmosphere rather than the true sizes of the stars, which are much smaller than this. Until recently, astronomers had only ever seen the disk of one star – the one that we see every day. The Hubble Space Telescope however is above the atmosphere, and Betelgeuse is so large that astronomers have recently succeeded in using the HST to obtain an image of its disk (Figure 4.3).

In 1952, Tycho's star returned to human history. In that year astronomers used one of the first radio-telescopes to detect radio

FIGURE 4.3 The first image of a star's disk apart from the Sun. The right-hand picture shows the constellation Orion; Betelgeuse is the star at the top left. The left-hand picture shows a picture of Betelgeuse taken with the Hubble Space Telescope. Even with the HST's ability to see fine detail, it is still only just possible to make out the disk. Credit: Andrea Dupree, Ronald Gilliand and NASA

waves from close to where Tycho had seen the star. A few years later, using a more traditional optical telescope, astronomers discovered some faint "knots" of light close to the position of the radio source. Using spectroscopy (Chapter 3), they discovered that the knots have such huge Doppler shifts that they must be travelling at speeds of up to 10,000 kilometers per second. The knots are all moving away from a central point, and the astronomers calculated that an explosion must have occurred some time towards the end of the sixteenth century. The new stars observed by Tycho and Kepler were not stars being born but colossal explosions – the ends of stars rather than the beginnings.

Why do stars explode? To see why this happens, we must return to the story of the Sun. The Sun will swell up to become a red giant, but eventually it will stop expanding. The reason for this is once again something that happens deep under its surface. While the Sun is on the main sequence, energy is generated by the fusion of hydrogen nuclei into helium nuclei. After the hydrogen is exhausted, the core, which will now mostly be helium "ash," will begin to contract; its temperature will begin to increase as gravitational energy is released. At a high enough temperature the helium ash will itself become useful as a fuel – the helium nuclei will start fusing to form carbon and oxygen nuclei. There will be a similar discrepancy between the mass of the ingredients and the mass of the product, and energy will be released. Once helium-burning starts, the Sun will stop expanding and will enter a new period of stability. This period will last another billion years.

By the end of the billion years, however, all the helium will have been used up. The core will be choked with new ash – carbon and oxygen nuclei – and nuclear fusion will stop. The same things will happen as when the hydrogen ran out. The core will slowly contract; its temperature will increase; the heat from the core will cause the outer layers of the star to expand – this time the next planet out from the Sun, Mars, will be vaporized. When a star of the Sun's mass gets to this stage, it becomes unstable. Material will be ejected from the Sun because of pulsations in the Sun's outer layers. The death of the Sun will produce one of the most beautiful objects in astronomy.

Figure 4.4 shows pictures taken with the Hubble Space Telescope of some stars to which these events have already happened.

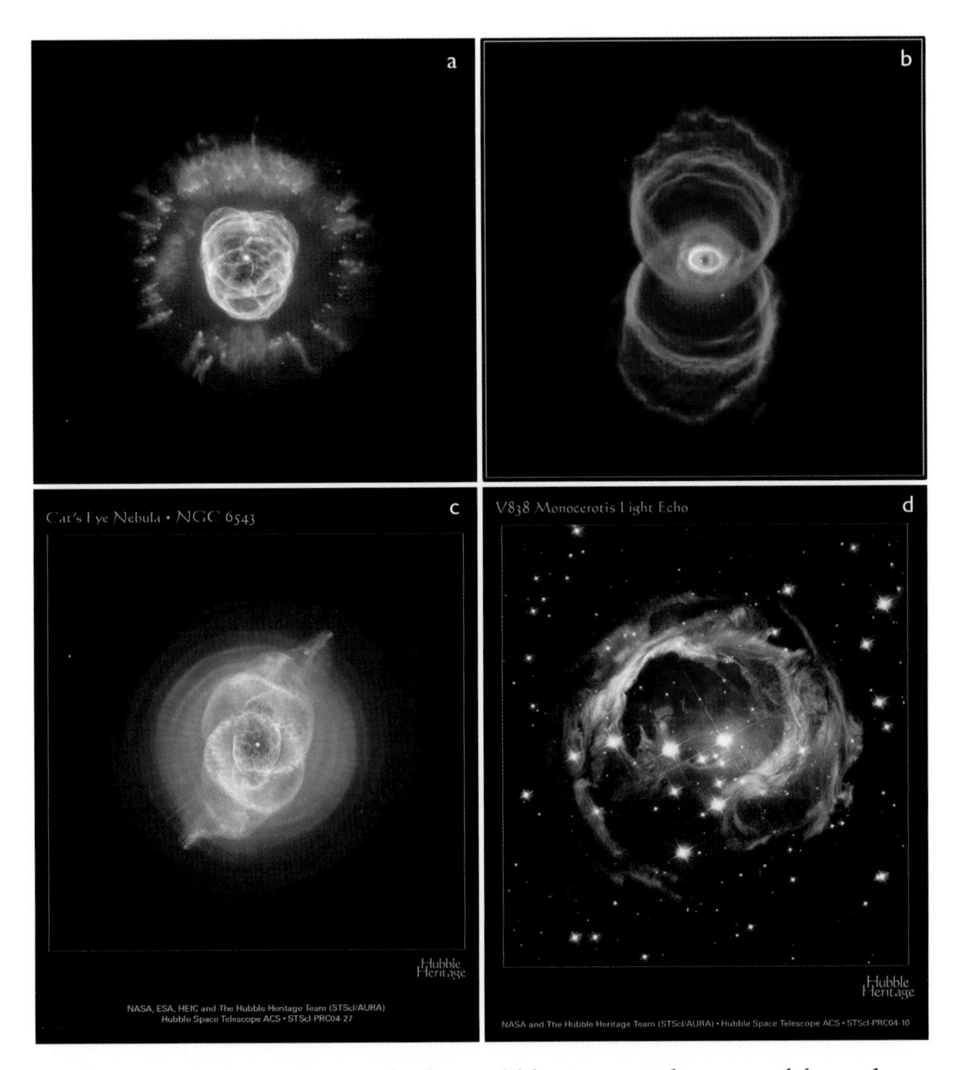

FIGURE 4.4 Images taken with the Hubble Space Telescope of four planetary nebulae. The planetary nebula at the bottom right still has a red giant at its center. Credits: The two lower images: The Hubble Heritage team and NASA. The image at the top right (the Hourglass Nebula): Raghvendra Sahai, John Trauger, the WFPC2 team and NASA. The image at the top left (the Eskimo Nebula): Andrew Fruchter, the ERO team and NASA

These are *planetary nebulae,* a completely misleading name, because these pictures have nothing to do with planets and show the deaths of stars. The beauty of the pictures lies in the complex structures of the ejected outer layers of the stars, which are now strewn over many light years (astronomers do not yet have a convincing explanation of these structures). The star at the bottom right is not yet dead – although it has clearly received the last rites – because it is still possible to see the red giant at the center of the nebula. For the other stars, the life cycle is complete, because all that is visible at the center of each nebula is a blueish-white point of light – the white-hot core of the star, now completely exposed by the removal of the star's outer layers. This is a white dwarf.

A white dwarf is a dead star, because although the star continues to shine, no new energy is produced, either by nuclear fusion or by the release of gravitational energy, inside the star. The energy lost by radiation from the star's surface is therefore not replaced, and as the star's reservoir of energy is depleted, it will gradually fade from view.

As the white dwarf's energy leaks into space, the temperature of the gas inside it will decrease, which should also cause its pressure to decrease. It is the pressure of the gas that stops a star collapsing, and so one might expect the decrease in pressure would make the star suddenly collapse. The reason the star does not collapse is that the gas inside the white dwarf is now so dense that an entirely new type of pressure comes into play. The pressure that holds up the weight of the Sun is exactly the same kind of pressure that holds up the weight of a car, which makes explaining it fairly easy. Unfortunately, there is no everyday example of this new type of pressure.

Quantum mechanics is the theory that describes the world of subatomic particles: protons, neutrons and electrons. The best way I can think of explaining quantum mechanics is to start by explaining what an electron is *not.* An electron is not a ball or a little planet orbiting a nucleus made up of other little balls. The picture of an atom as a mini-Solar System is an attractive one because it is based on concepts with which we are all familiar. At the beginning of the twentieth century, scientists discovered that sometimes an electron behaves like a wave. This discovery was not too disturbing, because again a *wave* is a familiar concept from our everyday world; the image of waves crashing against an ocean shore immediately springs to mind. The correct way however to think about the subatomic world is this: an electron is *not* a wave; it is *not* a little ball; it is not even a combination of the two. An electron, according to quantum mechanics, is something that cannot be described using familiar images from our everyday world – it is literally beyond our imagination. Of course, any theory constructed by human scientists must start with standard human concepts. The group of European scientists who constructed quantum mechanics at the beginning of the twentieth century started with the basic ideas of balls and waves. But the mathematical theory that they constructed on these foundations ended up by subverting the original ideas. An electron sometimes behaves as if it is a wave and it sometime behaves as if it is a little ball – and quantum mechanics allows scientists to predict how it will behave – but it is neither a wave nor a little ball nor a combination of the two; it is something *other*, something outside the realm of everyday ideas and images.

The pressure in a white dwarf arises because of a combination of an electron's wave-like and particle-like properties. According to quantum mechanics, it is only possible to squeeze a certain number of electrons into a fixed space. In the Sun's core at the moment there is plenty of room, but by the time the core has evolved into a white dwarf star, it will be much, much smaller; it will be so dense that a teaspoon of material scooped up from the star would contain about the same mass as a large truck. The electrons will therefore be squeezed into a much smaller space, and it is their resistance to being squeezed that will produce the strange pressure – the technical term is *electron degeneracy pressure* – that stops the star collapsing.

According to the theory of stellar evolution, a star with a fairly low mass like the Sun will end its life as a white dwarf. Some stars however end their lives in a much more spectacular fashion. A *supernova* may for a few months outshine a whole galaxy of one billion stars. By the standard of these colossal explosions, the Sun will end its life, as it shucks off its outer layers, in a series of cosmic burps.

A complication in explaining why supernovae occur is that we now know there are several types of supernova. Astronomers can fairly easily classify a supernova today by obtaining its spectrum, but Tycho's and Kepler's supernovae were seen before the invention of the telescope. Astronomers are fairly certain, from the account in Tycho's book, that his supernova was a Type 1a supernova. The class of Kepler's supernova is a bit more uncertain, but there is some evidence that it may have been a Type Ib supernova[13]. If these classifications are correct, the stories of Tycho's and Kepler's new stars were rather different.

The story of Tycho's new star began about 6000 BC in the Perseus spiral arm of our Galaxy about 7600 light years from Earth. The characters in this story were two stars in a binary star system. Binary systems are not unusual – roughly 60% of stars are in binaries – but this binary system was a little unusual because both stars, a red giant and a white dwarf, were close to the ends of their lives. The binary system was also unusual because of the small distance between the stars. Because they were so close and because the gravitational field of the white dwarf was much stronger than the weak gravitational field at the surface of the red giant, gas was being sucked across the small gap between the two. There were no nuclear reactions occurring in the white dwarf, which was mostly made of carbon and oxygen, but the material accumulating on its surface was hydrogen from the outer layers of the red giant – fresh nuclear fuel. The story is very brief, because astronomers do not fully understand why it happened, but at some point this nuclear fuel exploded, ripping the two stars apart. Seven thousand and six hundred years later, this explosion was seen by a young Danish aristocrat in the sky over Herrevad Abbey.

The story of Kepler's new star is a longer one, but it is one that is better understood by astronomers. The main character in this story was a star with a mass much higher than the Sun, probably about 25 times greater. The star was also much more luminous. The relationship between the luminosity and mass of a star that is on the main sequence is not the obvious one. A star that has twice the mass (and thus twice the fuel) does not simply have twice the luminosity; it actually has approximately two-cubed or eight times the luminosity. The luminosity of the star in this story was therefore 25-cubed or roughly 16,000 times greater than the luminosity of the Sun. Because the star was using its fuel at such a prodigal rate, it would have had a much shorter life than the Sun. The Sun will spend ten billion years on the main sequence; the star in this story spent only seven million years there.

At the time this story begins, approximately 16,000 years ago (the remains of Kepler's supernova are 15,800 light years from the Earth), the star was already a red giant. It had reached the same stage in its life that the Sun will reach just before it loses its outer layers; there was no helium fuel left in the core, which was now composed mostly of carbon and oxygen ash. Nuclear reactions however had not shut down completely, because there were two thin shells surrounding the core: an inner shell in which helium nuclei were still fusing to form carbon and oxygen nuclei and an outer shell in which hydrogen was still fusing to form helium.

At this point, the stories of the star and the Sun diverge. Stars this massive are not so unstable. As in previous phases, once nuclear reactions had stopped in the core, it began to contract. The temperature started to increase because of the release of gravitational energy. At a high enough temperature the carbon ash became useful as a fuel and carbon nuclei began to fuse. The mass of the products, mainly neon and magnesium, was slightly less than the mass of the ingredients. The disappearance of mass, according to Einstein's famous law, liberates energy, and this new phase of nuclear burning stopped the core contracting any further.

This period of "carbon-burning" lasted for only 500 years. At the end of this time, all the carbon had been used up. The core began to contract and get hotter. When the temperature became high enough, neon burning started.

The story of the star now moves from the eternities of the night sky onto the timescale of our individual human stories. In one year, less time than it has taken me to write this book, all the neon was exhausted. The core started to contract; the temperature increased; oxygen burning started. Oxygen burning, which produces elements such as silicon and phosphorus, lasted 140 days.

After all the oxygen was used up, the core started to contract again. The temperature increased and eventually silicon burning started. Silicon burning produces the nuclei of fairly heavy elements such as iron. *All the silicon in the core was consumed in approximately 20 hours.*

The story of the star now becomes a faster-paced story than our individual stories. The difference between the ash in the core now and the ash at the end of previous cycles of nuclear burning was that iron is not a potential nuclear fuel. Two iron nuclei may fuse to form a bigger nucleus, but the nucleus that is formed has more mass than the sum of the masses of the iron nuclei; energy is required for this process rather than produced. At this time, the star had an onion-shell structure. Around the core, now mostly composed of iron, there was a shell in which silicon was still being burnt; outside this there were, successively, oxygen-burning, neon-burning, carbon-burning, helium-burning and hydrogen-burning shells. Surprisingly, the core itself did not immediately collapse, because it was so dense that its weight was supported by electron-degeneracy pressure, the strange pressure predicted by quantum mechanics that I described above.

Electron degeneracy pressure however can only support stars up to a certain critical mass. This critical mass, called the *Chandrasekhar limit* after the Indian physicist who discovered it, is approximately 1.4 times the mass of the Sun. Initially the iron core was safely below this limit, but new iron was being continually added on to the core from the silicon-burning shell above.

At some time in the 15th millennium BC, the core of the star exceeded the Chandrasekhar limit. The core collapsed in less than a second. As the core collapsed through its own gravitational field, a huge amount of gravitational energy was suddenly released. The energy produced in one second was approximately one hundred times greater than the energy that will be produced by nuclear fusion in the Sun over its entire life of ten billion years. Most of this energy was carried away from the site of the collapse by a stream of tiny particles called neutrinos, particles that are so insubstantial that they pass through the Earth almost as if it is not there. Some of this energy however produced a huge explosion in the outer layers of the star, which led to a burst of nuclear fusion reactions – reactions which produced the elements heavier than iron, reactions that require energy rather than produce energy.

This was the explosion seen by Johannes Kepler in AD 1604.

The star collapsed to form either a neutron star or a black hole. A neutron star is a star that is prevented from collapsing by *neutron degeneracy pressure*, the same as electron degeneracy pressure but applied to neutrons rather than electrons. The difference between the two is that neutrons only exert this bizarre pressure if squeezed into an incredibly small space. If the core of Kepler's star did collapse to form a neutron star, its size must have decreased from 20,000 kilometers to 20 km in less than a second. Before the collapse the core was already extremely dense; after the collapse a teaspoon of material scooped up from the neutron star would have contained about the same mass as a large mountain. We know that neutron stars do form in this way, because some neutron stars emit pulses of radio waves, and these pulses have been detected at the positions where supernovae have occurred.

The other possibility is that the collapse never stopped. There is a maximum weight that can be supported by neutron degeneracy pressure, and if this weight is exceeded there is nothing left to halt a star's collapse. It is possible that the core of Kepler's star collapsed to form a *black hole*, a singularity in space-time (Chapter 9) from which not even light can escape. We do not know whether a black hole or a neutron star was formed when the core of Kepler's star collapsed. Nothing has been detected at the center of the cloud of glowing gas surrounding the site of the supernova (Figure 4.5), but this could be explained either way because neutron stars are also very difficult to detect and do not always produce pulses of radio emission.

The connection that Tycho tried to make in his book between the new star of 1572 and events on Earth does not exist. Stars are merely huge balls of gas, not signposts in the sky, and the new star was a supernova that did not foretell anything about human history (it actually happened eight thousand years ago). I do not know how many people nowadays believe in astrology. I could speculate wildly that either our culture is so inculcated with the ideas of science that nobody really believes any longer that the stars hold any meaning, or alternatively that we are currently being swamped by a tidal wave of New Age mysticism. However, as a scientist, one should always try to measure something. Searching *amazon.com*, I found 16,932 books on the subject of astronomy and 7474 on the subject of astrology. Thus in five hundred years the proportion of books about astronomy and astrology has reversed from 10-to-1 in favor of astrology to slightly over 2-to-1 in favor of astronomy.

But nevertheless there *is* a connection. Astronomers have used spectroscopy to show that the clouds of hot gas at the sites of supernovae (Figure 4.5) contain large amounts of elements such as carbon, oxygen and iron. These observations demonstrate conclusively that most of the common elements were made in the nuclear furnace at the center of a massive star.

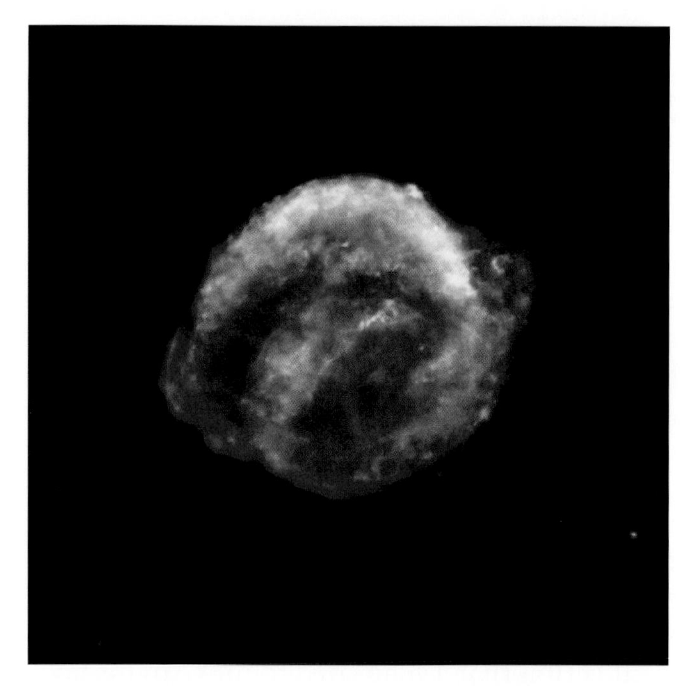

FIGURE 4.5 Image made with the Chandra X-ray observatory of the hot gas at the position of Kepler's supernova. Credit: NASA/JHU R. Sankrit and W. Blair

Almost everything on Earth is made of these elements. The Earth itself is composed mostly of silicon, magnesium, carbon, iron and oxygen – elements that would have been formed in the period leading up to a supernova. Without supernovae to scatter these elements throughout the Galaxy, our planet would not exist. As I type these words, I can see my wedding ring – the gold, an element heavier than iron, must have been formed in the conflagration of the supernova explosion itself. Our bodies, too, are almost entirely composed of elements that were formed in a star's nuclear furnace. The carbon and oxygen atoms in my body were created in the dance between helium nuclei deep inside a star. In the words of the songwriter Joni Mitchell: "We are stardust, billion-year-old carbon." This is the fundamental connection between supernovae and life on Earth.

5. The Final Frontier

The true birth of a star is much less dramatic than the sudden appearance in the sky of Tycho Brahe's new star. It is actually a furtive secretive affair, which astronomers have only recently become able to observe directly. Nonetheless, a genuine stellar birthplace is one of the landmarks of the night sky.

The constellation Orion is one of two that any professional astronomer, even a cosmologist, can be guaranteed to find*. The constellation, which is named after the hunter Orion in Greek mythology, is also one of the few constellations that look something like the character it is supposed to represent. With four bright stars to represent his hands and feet, and a chain of three bright stars to represent his belt, the constellation does at least look a bit like a human figure. It is also possible to see just below the belt a faint fuzzy patch of light, with the hint of something brighter in the center of the patch. If one is thinking in mythological terms, one might imagine that this fuzzy patch is Orion's sword; to astronomers, it is the Orion Nebula, a crowded stellar nursery.

Figure 5.1 shows a picture of the Orion Nebula taken with one of the big optical telescopes at La Palma Observatory. The star at the center of the nebula that can just about be seen with the naked eye is actually four stars in a small group. The patches of light in the picture are clouds of glowing gas. The radiation from the stars in the group, which is called the *Trapezium*, excites the

* It is always a disappointment to the friends and relatives of professional astronomers that many of us can not find our way around the night sky. The other constellation that most of us can find is the Big Dipper or Plough.

FIGURE 5.1 Optical image of the Orion Nebula. The Trapezium stars are just to the left and below the center of the picture. Credit: Image taken by Nik Szymanek, courtesy of the Isaac Newton Group of Telescopes, La Palma

gas, which then emits visible light – in somewhat the same way that the gas in a fluorescent light bulb is excited by electricity passing through it. The dark patches in the picture I will explain in a moment.

The first piece of evidence that stars are forming in the Orion Nebula is circumstantial. As I explained in the last chapter, stars with higher masses than the Sun, even though they have more fuel to begin with, use up this fuel at a much faster rate. The brightest star in the Trapezium is the star at the bottom of the group, which has the eminently forgettable name of Θ^1 Orion C. This star's mass is about 30 times that of the Sun, but its luminosity is 300,000 times greater. Θ^1 Orion C has therefore 30 times as much fuel as the Sun, but it is burning this fuel at such a prodigious rate that it will have a total life of only about one million years. (This means incidentally that the night sky seen by our ancestors in Africa was not quite the same as the night sky we see today.) After stars are born, they tend like humans to drift away from their birthplace, but Θ^1 Orion C is such a young star that its birthplace must be very close to its present position.

The second piece of evidence is that stars have to form out of something. The Orion Nebula contains a reservoir of gas, and so it seems plausible that stars might be forming there. The mass of hot gas is actually not particularly large, but there is a lot more to the nebula than meets the eye.

The photograph in Figure 5.1 was taken with a large optical telescope, but most of the progress in astronomy during the last fifty years has not come from beautiful pictures like this taken with traditional optical telescopes, but through observations with completely new types of telescope.

What we refer to as light actually consists of electromagnetic waves with wavelengths between 300 and 900 nanometers (one meter contains a billion nanometers). This wavelength range is a tiny part of the complete electromagnetic spectrum, which extends from gamma-rays, which have wavelengths one million times less than visible light, to radio waves which may have wavelengths of over one meter (Figure 5.2). Our eyes are sensitive to this minuscule fraction of the electromagnetic spectrum for the very good reasons that the Earth's atmosphere is transparent to visible light and that the Sun emits most of its energy in this waveband; it would not make much sense if evolution had given us eyes that are sensitive to x-rays. However, when astronomers started to

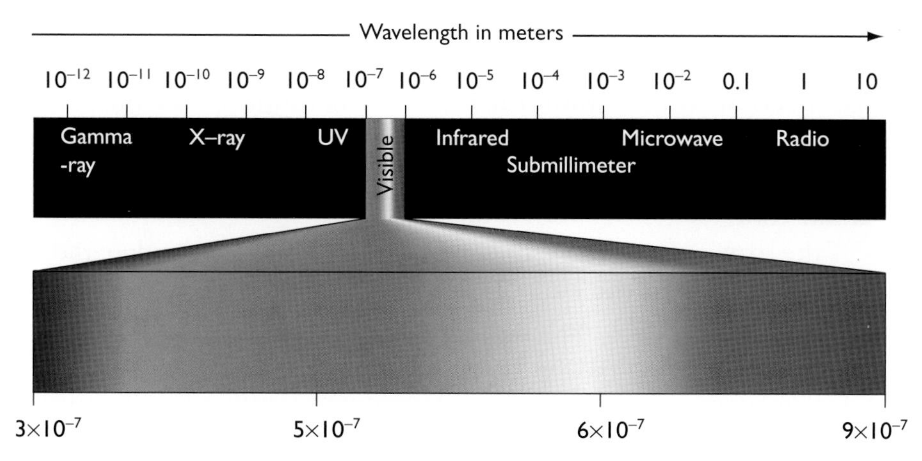

FIGURE 5.2 The electromagnetic spectrum.

construct telescopes to observe in the other electromagnetic wavebands, they discovered that our eyes give us a very biased view of the Universe.

The first of these was the radio waveband. Radio-astronomy got started properly just after the Second World War, when the scientists who worked on radar during the war came home and wanted to use their knowledge for a peaceful purpose. During the last fifty years, radio-astronomers have discovered phenomena which were unknown before the construction of radio-telescopes, including quasars, pulsars and the existence of an ocean of radiation filling the Universe today (including the room in which you are reading this book) which was produced shortly after the Big Bang (Chapter 8). The radio waveband was the first electromagnetic frontier to be opened up for astronomy because the technology was suddenly available after the war and because the Earth's atmosphere is transparent to radio waves; other types of astronomy, such as ultraviolet astronomy and x-ray astronomy, which are not possible from the ground because the Earth's atmosphere is opaque at these wavelengths, had to wait for spaceflight. During the last fifty years, every time astronomy pioneers have opened up a new electromagnetic frontier, they have seen many new things and obtained a new perspective on the Universe.

Until recently, astronomers tended to stick to the waveband they knew. Somebody who did a Ph.D. in radio-astronomy, for example, would generally stay a radio-astronomer for life. This was partly because of natural human inertia, but also because a new set of specialized techniques have to be learned every time one moves to a different waveband. During the last two decades, it has become more common to "waveband-hop," as observers have realized that if they want to understand some phenomenon – quasars, for example – it is important to use every possible observational tool.

Take my own case. I did a Ph.D. in the Cambridge radio-astronomy group in the early 1980s. I did dabble in optical astronomy during my Ph.D. because I wanted to obtain images of the galaxies that were emitting the radio waves I was detecting with the Cambridge radio-telescopes. However, I always felt mildly embarrassed by this. The group at the time contained many of the giants of radio astronomy, scientists who had built the first radio-telescopes, and I definitely got the impression from them that a real astronomer was someone who designed, built and used his own telescope; someone who wanted to flit between wavebands was clearly, in their eyes, a dilettante. When I moved to Hawaii, I went to work for two infrared astronomers. As they were prepared to pay me, it seemed only polite to make some infrared observations. Since that time, I have made x-ray and submillimeter observations, and the only wavebands that are missing from my observer's album are the gamma-ray and ultraviolet wavebands. Any astronomer who is my age or younger could probably tell a similar story.

According to our eyes, the space between the stars in the night sky is empty, apart from the occasional small and hard-to-see patch of gas such as the Orion Nebula. According to radio-telescopes, it is chock full of stuff. In 1970 radio astronomers detected a spectral line with a wavelength of 2.6 millimeters. This spectral line is produced by carbon monoxide, a molecule comprising one carbon atom and one oxygen atom. Over the last three decades, radio-astronomers have detected carbon monoxide at many places in the Galaxy. Carbon monoxide is not very interesting in itself, but carbon and oxygen are not particularly common elements, and astronomers have calculated that for every carbon monoxide molecule in interstellar space there must be approximately ten thousand molecules of hydrogen (two atoms of hydrogen stuck together) which cannot be detected directly. The Galaxy does not therefore just contain stars – it also contains a large amount of molecular gas. This new branch of astronomy, molecular-line astronomy, has revealed a hidden Universe, one that was completely unsuspected by traditional optical astronomers. One hundred and twenty-three different chemical compounds, from simple molecules like carbon monoxide to more complicated molecules such as alcohol, have now been detected in interstellar space.

This gas is not smoothly distributed through space, but is mostly contained in huge clouds called giant molecular clouds. The Orion Nebula is a small part of the Orion Molecular Cloud (Figure 5.3), a cloud containing enough gas to make one hundred thousand stars of the mass of the Sun. The hot glowing gas visible in Figure 5.1 is the tiny fraction of the gas, about one hundredth of one per cent, that is hot enough, as the result of the radiation

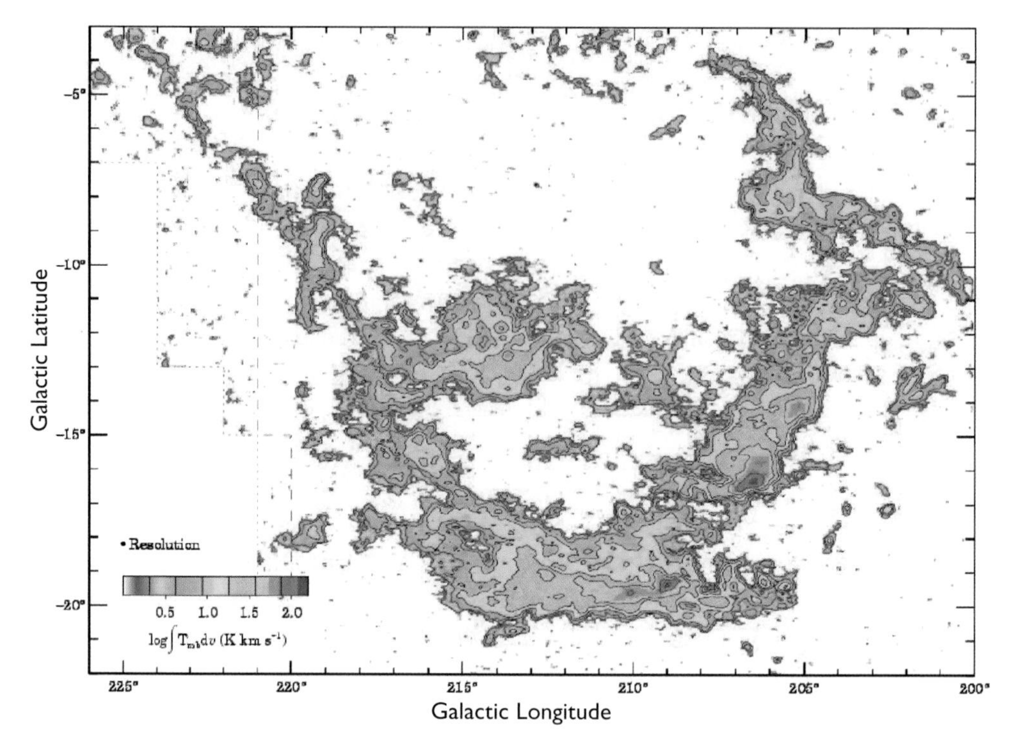

FIGURE 5.3 Map of the Orion Molecular Cloud made using observations of the carbon-monoxide spectral line.

from Θ^1 Orion C and the other Trapezium stars, to be emitting visible light. The evidence that stars form in giant molecular clouds is circumstantial but overwhelming. Stars have to form out of something; giant molecular clouds are the only significant reservoir of dense gas in the Galaxy; and the clinching evidence is that young stars, such as Θ^1 Orion C, are found close to giant molecular clouds.

Nevertheless, although Θ^1 Orion C is a young star in comparison to the Sun, it is at least a toddler – it is *not* a newborn star. The ideal way to investigate how stars are formed would be to watch a birth. Unfortunately, it is not possible to do this even with the largest optical telescopes, and for an astronomer who is interested in understanding how stars are formed, traditional optical astronomy is a dead end. The reason are the dark patches in Figure 5.1.

These patches are not places where there are no stars. They are places where our view of the stars is being hidden. The thing hiding the view is interstellar dust – tiny solid particles that along with the gas fill interstellar space. Not much is known about these dust grains, except that they are made out of the same sort of heavy elements that make up the Earth – carbon, oxygen, silicon, magnesium and so on. One thing that *is* known is that the dust grains scatter and absorb light. A better name for interstellar dust is interstellar smoke, because smoke also consists of tiny solid particles and also veils the light from anything behind it. The size of the particles in cigarette smoke is also typical of the size of many interstellar dust grains, although these probably range in size down to particles that are not much more than large molecules and up to – well, we are not sure if there is an upper limit (arguably, our planet is simply a large dust grain). Dust is responsible for some of the prettiest pictures in astronomy (Figure 5.4). However, from an astronomer's point of view, dust is a royal pain-in-the-neck because it obscures our view of the Universe.

Dust is even more of a pain-in-the-neck for an astronomer who is interested in watching the birth of a star, for a very simple reason. Gas and dust are always intermingled, and so when a gas

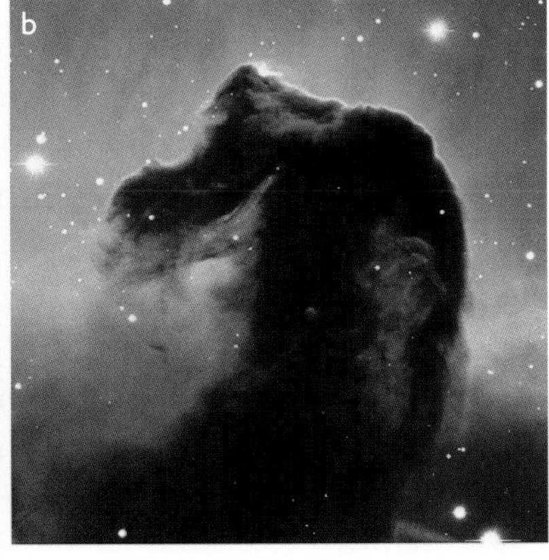

FIGURE 5.4 Two of the prettiest pictures in astronomy. On the left is the Eagle Nebula and on the right is the Horsehead Nebula. Both are stellar nurseries. The forms in the picture are caused by dust, which hides the light of the background stars. Credits: Image on the left: J. Hester, P. Scowen and NASA; image on the right: ESO

cloud begins to contract to form a star, the density of the gas and the dust will both begin to rise. Just at the moment things begin to get interesting, the dust will discretely veil the birth from sight.

The birth of a star is therefore a private affair, hidden from snooping by optical astronomers. But there is one way of telling that something is happening inside a molecular cloud. Except in the alchemist's furnace in the middle of stars (Chapter 4), energy is never lost or created – it is simply converted from one kind into another (from gravitational energy to kinetic energy to heat, in the case of a collapsing star or a diver diving from a high board; from chemical energy to heat to kinetic energy, in the case of a car engine). The energy in the visible light that is absorbed by the dust is not lost, but instead warms up the dust. The radiation from this warm dust is a signal that a star is being formed deep inside a cloud. It is fairly simple to show (I will explain how below) that the wavelength of this radiation should be in the infrared waveband. In the 1960s, when astronomers began to develop the techniques and instruments to detect this radiation, they were driven by a single scientific goal. In the words of one of the pioneers of infrared astronomy, Gareth Wynn-Williams, who incidentally was one of my bosses in Hawaii: "Protostars are the Holy Grail of Infrared Astronomy."[14]

The reason that infrared astronomy did not start until the 1960s is that infrared radiation is also emitted by everything on the Earth. We are unaware of the flood of infrared photons that streams from everything around us because the light-sensitive cells at the backs of our eyes are only sensitive to visible light; I see the cars and trees outside in the street because light from the Sun is reflected by them and then enters my eyes. It is blindingly obvious that the Sun *emits* radiation, but every other object in the Universe – the planets, the dust in interstellar space, even our bodies – also emits electromagnetic radiation. It is just that we can't usually see it.

The wavelength of the radiation that is emitted by an object depends on its temperature. Scientists generally use the Kelvin temperature scale rather than the everyday Centigrade scale, but the conversion between the two is very easy: the temperature in Kelvin is equal to the temperature in Centigrade plus 273 degrees. The Kelvin scale was defined to reflect the fact that the coldest temperature possible, when all energy has been extracted from an object, is –273 degrees Centigrade, which on the Kelvin scale is 0 degrees – *absolute zero*. With some moderately complicated theory from thermodynamics, the branch of physics that deals with temperature and energy flow, scientists can calculate the properties of the radiation emitted by an object at any temperature. The useful rule that comes out of the theory is that the characteristic wavelength of the radiation is inversely proportional to the object's temperature – if the temperature of the object increases by a factor of two, the characteristic wavelength falls by the same factor.

With this rule and two facts about the Sun, we can calculate the characteristic wavelength of the radiation emitted by an object at any temperature. The Sun emits most of its radiation in the visible waveband, which has a central wavelength of about 0.5 microns (one micron is a millionth of a meter). The temperature of its surface is about 6000 K. Astronomers have estimated that interstellar dust grains have a temperature of between 20 and 60 K, the precise value depending on how close the dust grain is to a heating source. Let us take the high value. A temperature of 60 K is a factor of 100 times less than the Sun's temperature, and so according to the useful rule I gave above the characteristic wavelength of the radiation from the dust is 100 times greater, which gives a wavelength of 50 microns. This is in the middle of the infrared waveband, which is therefore the waveband of choice for any astronomer interested in dust. The fundamental problem faced by infrared astronomers, however, can be seen by carrying out the same calculation for objects on the Earth. Our bodies, for example, have a temperature of about 300 K, which leads, using the same reasoning, to a characteristic wavelength of about 10 microns – also in the infrared waveband. This flood of infrared photons from objects on the Earth and from the Earth's atmosphere is the fundamental problem faced by infraredastronomers; their ritual complaint is that trying to detect infrared radiation from a faint star is like trying to detect the star's visible light but in the middle of the day and with a luminous telescope.

Nonetheless, by the 1960s astronomers had developed techniques to do this. Towards the end of the decade, astronomers detected infrared emission from two places in the Orion Nebula where there was nothing visible at optical wavelengths. In the paper announcing the discovery of one of these radiation sources, the American astronomers Eric Becklin and Gerry Neugebauer, two of the other pioneers of infrared astronomy (by a coincidence, Becklin was my other boss in Hawaii), remarked

> It is well-known that the Orion Nebula is a very young association and that the probability of finding a star in the process of forming should be relatively high ... Thus an attractive interpretation of the observation is that the infrared object is a protostar.

Astronomers are human beings and naturally put the best spin on their results, but at the time the evidence for this conclusion was not very watertight – in fact it leaked like a sieve. In particular, there was the obvious possibility that the object hiding behind the dust was simply a mature star on the main sequence. The term Becklin and Neugebauer used for a newborn star, a *protostar*, was also very vague (arguably, the history of this subject over the last four decades can be summed up as an increasingly precise definition of what we mean by the term). The importance of their observations and those of the other infrared pioneers was that they showed there are energy sources in the Universe that are completely hidden by dust from traditional optical telescopes.

Discovering a hidden energy source is therefore not quite the same thing as discovering a newborn star. To see what signs we should look for that will tell us there *is* a star being born within a molecular cloud, we need to consider how theorists think stars are formed.

It seems likely that stars are formed in the densest parts of molecular clouds, which are called *molecular cores*. The details of what happens during the collapse of a core to form a star are very uncertain (this, of course, is why this is one of the most vigorous areas of astronomical research). As the core contracts from a rarefied cloud of interstellar gas to a star – a change in density of approximately one billion billion – many different physical and chemical processes will occur at different stages. It is virtually impossible to make any progress without simulating these processes on a computer. This is also true of the formation of planets (Chapter 1), but there is one big additional problem for scientists trying to simulate the birth of a star.

Life would be much simpler for the modellers if they could be sure that a star is formed out of an isolated spherical cloud of gas that collapses under its own weight. The big problem is that nobody is sure that a star really does form out of an isolated spherical cloud, and if this assumption about the *initial conditions* of the simulation is wrong, the results of the simulation will be worthless (the GIGO problem – *garbage in, garbage out*). Moreover, observers have already found evidence from radio observations of molecular clouds that this assumption is probably wrong. The clouds have a remarkable filamentary structure (Figure 5.3), which has the interesting property that observers see the same kind of structure whether they observe a cloud with a small radio-telescope, which allows them to obtain only a broad-brush picture of the gas, or whether they use a large radio-telescope to study the fine detail of a small region of the molecular cloud. A plausible way of producing this filamentary structure that looks quite similar on all scales – the technical term is *self-similar* – is as a result of shock waves from supernovae sweeping through the interstellar medium, which is an interesting connection between the births and deaths of stars. The complexity of the structure and the turbulence of the interstellar gas, shown by the Doppler shifts of the spectral lines, suggest a variety of possible ways cores might be formed and then be tipped into collapse.

Figure 5.5 shows a result from one simulation, which I have only used because the offices of the modellers are just down the corridor from me in Cardiff. As the initial conditions for their simulation, the modellers decided to start with two gas clouds that are about to collide. These may be quite realistic initial conditions, because the turbulence observed in the interstellar medium implies that collisions between gas clouds should be quite common. The simulation shows that when the gas clouds do collide, a layer of gas between them is compressed and heated by shock waves. The gas in this layer then cools and fragments into separate molecular cores, which then collapse to form stars. A nice result of this simulation is that many of the stars end up in binary star systems, which agrees with the observation that roughly 60% of stars are in binaries.

The modellers however could have started with different initial conditions. They could have chosen different masses for the

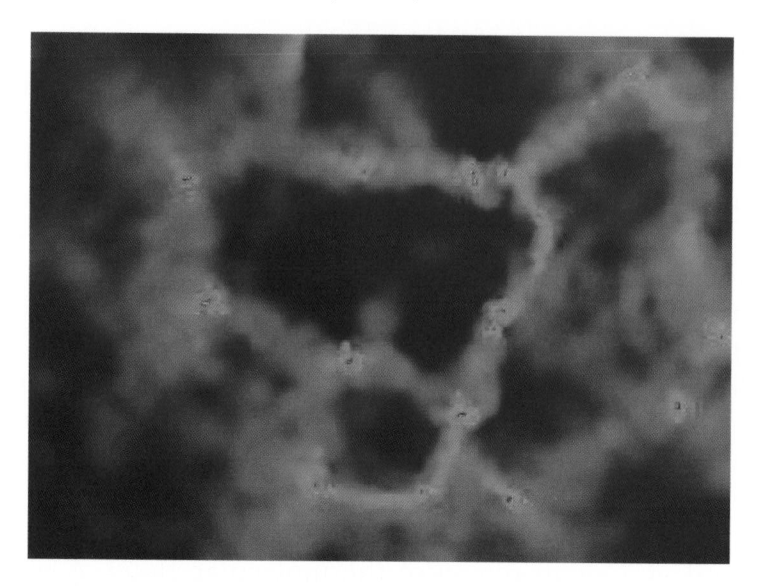

FIGURE 5.5 A simulation that starts with the collision of two gas clouds. The picture shows part of the layer of shocked gas created by the collision. The dark specks in the image are stars which have been formed in the shocked gas. Credit: Ant Whitworth

clouds or fired the clouds at each other at a different speed. Because molecular clouds contain filaments, the modellers might also have started their simulation with two colliding filaments, opening up a whole range of possibilities for the choice of the properties of these filaments. In this simulation, shock waves are produced by the collision of the two gas clouds, but the modellers might have started their simulation with shock waves from a supernova passing through a single cloud or filament. It is this smorgasbord of choices for the initial conditions that makes simulating the birth of stars so challenging. This simulation is also a nice illustration of the other problem faced by modellers. These modellers have chosen, because of the usual limitations of computer power, not to include magnetic fields in their simulation. Other theorists believe however that the magnetic field in the interstellar gas plays a critical role in the formation of stars.

The theorists however do all agree about some things that will happen when a star is formed. As gas falls towards the center of the collapsing molecular core, it is falling through a gravitational field, and so the usual transformation of energy will occur:

first, gravitational energy into kinetic energy, the energy of motion; and then, as bits of gas crash into other bits of gas, kinetic energy into heat. The temperature of the gas will begin to rise, and the pressure in the gas will also rise because the pressure of a gas depends on its temperature (molecules in a hot gas move more quickly and therefore have more oompf when they hit something). The first moment in the life of a star is debatable, but possibly it is when the pressure at the center of the molecular core becomes high enough to stop the gas there collapsing any further. What happens now, according to the theorists, is that a kernel will form within the molecular core. The pressure in the kernel will be high enough to support its own weight and stop it collapsing any further; outside the kernel, gas will continue to fall inwards, raining down onto the kernel's surface. This kernel is not really a star, of course – its density, pressure, temperature and mass are much too low for nuclear fusion to have started – but arguably, this is the moment of conception.

The theorists predict that an object like this should be surprisingly luminous. In the gas raining down onto the surface of the kernel there will be the usual transformation of gravitational energy into heat. The hot gas will emit radiation and, according to the theorists, the amount of radiation emitted by the gas may be as much as that produced by a fully grown star like the Sun, in which nuclear reactions are already going full blast. The dust absorbs any visible light from the gas, but the warm dust will emit radiation at infrared wavelengths. The "protostar" detected by Becklin and Neugebauer is pretty much what the theorists predict one should see: a luminous infrared source deep in a molecular cloud. As with money-laundering, the big problem is that the dust conceals the original source of the energy. One can tell how much energy is being produced, but one can not tell whether the ultimate source of the energy is nuclear fusion in the center of a star or the drizzle of gas onto the surface of a protostar.

By the 1980s, however, there was convincing evidence that many of the sources detected by infrared astronomers *are* protostars. There were actually two pieces of evidence. The first was that the hot gas in a protostar is expected, for reasons I have not space to go into, to also emit radio waves; by the early 1980s, faint radio sources had been detected at the positions of many of

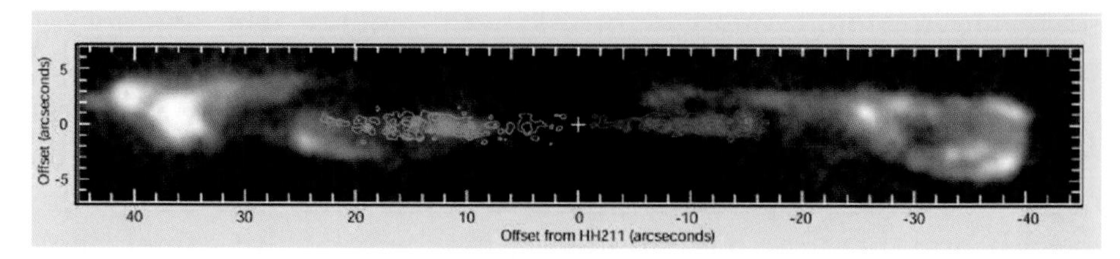

FIGURE 5.6 A molecular jet from a protostar. The cross marks the position of the protostar.

the infrared sources. The second was the discovery of molecular jets.

The existence of molecular jets was completely unsuspected. The obvious way to test whether a molecular core is collapsing is to look for the Doppler shifts produced by the in-falling gas. When radio-astronomers started to look for these shifts, they found to their surprise that they often saw the opposite: a huge jet of gas heading *away* from the position of the infrared source (Figure 5.6). We still do not have a very good understanding of the cause of a *molecular jet*, but theorists believe it is connected to the formation of a disk around a protostar.

By the late 1980s, infrared astronomers could therefore congratulate themselves that they had found their Holy Grail. Overcoming huge technical obstacles – optical astronomers do not have to use telescopes that emit visible light – they had finally discovered newborn stars. The infrared astronomer Charles Lada even invented a classification system for protostars. In Lada's system, Class II and Class III protostars are objects in which nuclear fusion has already started but which have not yet reached the main sequence. Protostars in both classes can be seen with optical telescopes. In this system, an object in Class II is slightly younger than one in Class III, because although the protostar is visible, it is still surrounded by some dust, showing it has not yet moved out of the stellar maternity ward. Arguably, of course, both Class II and Class III objects are not really protostars at all, because nuclear fusion has already started. A Class I object however is a protostar by any definition. This is an object which is so hidden by dust that it can only be detected by infrared astronomers. In these objects, which often have molecular jets, the ultimate energy source is almost certainly not nuclear fusion but instead the conversion of gravitational energy into heat. At the end of the decade, infrared astronomers could reasonably claim that they had identified the first moment in the life of a star.

A few years later, three astronomers showed they were wrong. By this time, the pioneers of infrared astronomy were all middle aged. As is often the way in science, the astronomers who showed they were wrong were very young. One was American, one was French and one was British.

The American was Mary Barsony. In the early 1990s, Mary was a postdoc at Harvard, having just finished a Ph.D. at the California Institute of Technology. During her Ph.D. she had used one of the radio-telescopes at the Owens Valley Radio Observatory to observe the molecular jets from some well-known protostars. However, in the time of reflection that follows the hustle and bustle of finishing a Ph.D., she began to wonder whether the claim of infrared astronomers that a Class I protostar is the earliest moment in the life of a star was actually correct. Her reasoning was quite simple. The dust close to the center of a protostar emits infrared radiation. Unlike visible light, the infrared radiation passes through any dust that is further away from the protostar relatively unscathed – very little of it is absorbed by the dust. Very little, but not none. Mary realized that the dust around some protostars, especially extremely young ones, might be so dense that even infrared radiation would not be able to escape.

At star-formation conferences at the end of the 1980s and beginning of the 1990s, Mary began to suggest that, just possibly, there might be a younger kind of protostar – one that does not show up in infrared surveys. This can not have made her very popular with infrared astronomers. If one comes back victorious from a twenty-year quest, having found the Holy Grail, it must be slightly irritating to be told: well, no, that is not quite what we wanted, go out and try again.

The Frenchman was Philippe André. In the early 1990s, Philippe was a postdoc at the National Radio Astronomy Observatory in Tucson. He also wanted to understand the origin of stars, and while he was working in America he had started to use the Very Large Array, the world's largest radio-telescope, to observe the radio waves in the direction of several molecular clouds. He discovered faint radio sources at the positions of many Class I protostars, which was not unexpected because theorists had predicted that protostars should emit radio waves. However, in one molecular cloud, ρ Ophiuchus A, he found a weak radio source at a position where there was no infrared radiation at all.

This was not necessarily interesting. The sky is full of radio sources, most of which are quasars and distant galaxies, and it was perfectly possible that the radio source in ρ Ophiuchus A was simply a quasar behind the molecular cloud*. However, Philippe happened to hear Mary give a talk at a conference in Granada in Spain, and he began to wonder whether this radio source, VLA 1623, might be one of Mary's highly obscured protostars.

Before describing what happened next, I have to take a short historical break and tell the story of the pioneers of the final electromagnetic frontier. As I described earlier, most of the progress in astronomy in the twentieth century was the result of the gradual opening up of the different electromagnetic wavebands for astronomy: the radio waveband after the Second World War; the infrared waveband in the 1960s; then, with the advent of space flight, the x-ray, ultraviolet and gamma-ray wavebands. The submillimeter waveband, which contains electromagnetic waves with wavelengths between 100 microns, roughly the thickness of a human hair, and one millimeter, was the last of these electromagnetic frontiers (Figure 5.2).

The reason for this were the obstacles faced by the pioneers. The first was the same as that faced by the pioneers in the infrared waveband. Everything on Earth emits submillimeter radiation. The infrared pioneers in the 1960s had been able to stop their detectors, at least, emitting infrared radiation by cooling them down to a temperature of 70 K. The huge obstacle in front of the pioneers in this new waveband was that even at 70 K a submillimeter detector is still emitting large numbers of submillimeter photons, which is the last thing you want if you are trying to detect a handful of photons from a faint star or galaxy. The only way to prevent a submillimeter detector from itself glowing with submillimeter

* Dust is completely transparent to radio waves and so it is possible to detect quasars even in the direction of a molecular cloud.

radiation is to cool it to a few degrees above absolute zero. It is relatively easy to cool a detector down to 70 K by using liquid nitrogen, but the only way to cool a detector down to the lower temperature is to use liquid helium, which is harder to work with and requires complicated and bulky cryogenic equipment.

The second obstacle faced by the submillimeter pioneers was that this kind of astronomy is only possible in certain special places. The Earth's atmosphere *does* let through submillimeter radiation, but only if the atmosphere contains hardly any water vapor. There are very few places on Earth dry enough for submillimeter observations to be possible at all. One of these is the 18,000 foot Atacama Desert in Chile, one is the 14,000 foot summit of Mauna Kea and one, strangely enough in view of the endless expanse of ice, is Antarctica*. All of these, needless to say, are not that easy to get to, and so the submillimeter pioneers were faced with the challenge of working in remote places with equipment that was hard enough to get working properly in a well-equipped laboratory at sea-level.

The final obstacle was that submillimeter signals from astronomical objects are very faint. The brightest source of submillimeter radiation outside the Solar System happens to be the Orion Molecular Cloud. Submillimeter detectors are essentially tiny thermometers which measure the heat received from an astronomical object (to feel the heat carried by radiation, hold up your hands towards the Sun on a hot summer's day), but the warming effect of submillimeter radiation is not exactly large. If you wanted to boil an egg using the energy received from the Orion Molecular Cloud by all the submillimeter telescopes in the world, you would have to wait about one hundred billion years[15].

Preston is not somewhere that springs to mind when one thinks of the main international centers of astronomy. It is a small Lancashire mill town, although most of the mills have now closed down. With its streets lined with endless terraces of mill-workers' houses, there is little to distinguish it from hundreds of similar small industrial towns in the English north and midlands.

* There is very little water vapor in the atmosphere; all the water in Antarctica is frozen.

Nevertheless, for a brief period in the 1980s and 1990s, Preston was one of the world centers of submillimeter astronomy, because it contained one of the few groups prepared to face the challenges of this final electromagnetic frontier (ironically, given the need of submillimeter astronomers for a very dry atmosphere, Preston has one of the highest annual rainfalls in England).

A small group of British astronomers, led by Peter Ade and Ian Robson, had built one of the first submillimeter detectors*. Ian Robson had subsequently moved to Preston where he gathered around him a group of young astronomers prepared to try to observe in this difficult waveband. Unfortunately, at the time there was no such thing as a submillimeter telescope, and so the Preston group and the handful of other submillimeter groups around the world were forced to use other peoples' telescopes. The Preston group used the United Kingdom Infrared Telescope (UKIRT) on the summit of Mauna Kea. They would take their detector out to Mauna Kea, mount it on UKIRT, and hope desperately for good weather (even on Mauna Kea, the atmosphere is dry enough only about fifty per cent of the time). After a few days, the infrared astronomers would ask for their telescope back, and the submillimeter astronomers would have to unbolt their detector from the telescope and go home. This was not ideal, but by the middle of the 1980s submillimeter astronomers had at least succeeded in detecting the planets, the Orion Molecular Cloud and some of the other brighter submillimeter sources in the Galaxy.

Eventually in 1987 the submillimeter astronomers got their own telescope. The James Clerk Maxwell Telescope is located on a small plain about two hundred feet below the summit of Mauna Kea, which has come to be called "Millimeter Valley." The JCMT was designed and built by a team of British scientists led by Richard Hills at Cambridge, but it is now jointly run by the UK, Canada and the Netherlands, with the University of Hawaii taking its small, irritating (for those not at the IFA) slice of the observing time.

---

* For any submillimeter wonk that ever reads this book, this instrument was UKT14.

By chance, I was the first astronomer to use the JCMT. At the time, this did not strike me as an important historical event and, to be honest, I was not that interested in submillimeter astronomy. As a young postdoc at the IFA in Hawaii, I spent my three years there trying frenetically to use as many different telescopes as possible, and the JCMT was only one among the crowd of telescopes on Mauna Kea. Moreover, its behavior during this first observing run did not exactly impress me with the possibilities of this new branch of astronomy. For an observer the minimum requirements of a telescope are that it point in the right direction and that it can track an object as it moves across the sky. During this inaugural observing run, the JCMT did neither, and for years I had on my office floor a pile of spectra of the galaxy M82 – spectra that were useless because I had no idea which parts of the galaxy I had been observing. This pile (possibly an important historical document) eventually vanished during one of my many moves.

The one disadvantage of building the first telescope for a new waveband is that there are usually very few astronomers with the experience necessary to use it. In the late 1980s, most of the astronomers in the UK with this experience were based in Preston, and indeed one of the first instruments mounted on the JCMT was the same one that the Preston astronomers had been schlepping out to the UKIRT. One of the astronomers then in Preston was the third member of the group that showed that infrared astronomers had not, after all, seen the first moment in the life of a star.

I have never met Mary Barsony and I know Philippe André only slightly. However, I know the third member of the trio, Derek Ward-Thompson, very well because his office is just across the corridor from me in Cardiff. The location of Derek's office is a bit of a problem, because one of his main characteristics is a spectacularly loud laugh, which for anyone with an office on the same floor can occasionally make it hard to work. As is only appropriate for someone born in the north of England, he also has a reputation for plain speaking; Derek tells it like it is, and if someone does not like what he says, he just says it again. In the early 1990s, after finishing a Ph.D. at the University of Durham, Derek was a postdoc in Preston.

After this short historical interlude, let me reprise the situation. Mary Barsony had suggested that the youngest protostars might be so heavily obscured by dust that even infrared radiation would not be able to escape through the dust. Philippe André had found a faint radio source in a molecular cloud at a position where there was no infrared radiation, which he suspected might be one of Mary's extremely young protostars. The new field of submillimeter astronomy offered a way of turning this suspicion into certainty, because submillimeter radiation passes through dust even more easily than infrared radiation. Therefore, even if the dust was so thick that infrared radiation from the vicinity of the protostar could not penetrate through the surrounding dust, it was possible that the submillimeter radiation might escape. If Philippe's suspicion was correct, at the position of the faint radio source, VLA 1623, there should also be a submillimeter source.

As it happened, Derek had already made submillimeter observations of this molecular cloud. This had been before the construction of the JCMT and he had used the Preston detector mounted on the UKIRT. He had detected submillimeter radiation in roughly the correct direction, but with a relatively small telescope such as the UKIRT it is possible to paint only a broad-brush picture of the sky, and he could not be sure the radiation was actually coming from the position of the radio source. The construction of the JCMT however opened up an interesting possibility, because with a big telescope such as the JCMT it is possible to observe more detail in the submillimeter sky and thus measure the position of a source with much more precision. Philippe knew of Derek's earlier observations, and in 1990 he wrote a letter to him, suggesting that they and Mary Barsony use the JCMT to test whether the submillimeter radiation was actually coming from VLA 1623. They wrote a proposal to use the JCMT; it was accepted; and in February 1991 the three young astronomers set off up Mauna Kea.

This was one of the few observing runs when everything went perfectly. Admittedly, they had been given the telescope's graveyard shift. The first shift of the night on the JCMT is rather civilized. You have dinner, drive up to the summit, and you are in bed by 2.30. On the graveyard shift, you have to hang around by yourself in the evening at the astronomers' residence when everyone else is at the telescopes; you drive up to the summit at one in the morning just at the moment your physical and mental resources are at their weakest; and you get back down to the residence at about ten in the morning for breakfast (if you have remembered to phone down from the summit to ask the kitchen staff to put a breakfast aside for you), by which time all the other astronomers have gone to bed. Nonetheless, the graveyard shift at the JCMT is one of the minor tribulations of life as a submillimeter astronomer, and I imagine that Derek and company reached the telescope (Figure 5.7) just before 1.30 in the morning on February 15th, 1991 in reasonably high spirits. Their moods would have become even sunnier when they discovered there was hardly any water vapor in the atmosphere and so the observing conditions would be excellent.

I would like to make the story of this observing run as dramatic as possible, and describe how over the three-day observing run the moods of the astronomers oscillated between exhilaration and depression, and that eventually, at the end of the run, tired and emotionally drained, they made their key discovery. Unfortunately, it was not like that. They made their discovery almost at once. After a quick observation of Uranus to calibrate the detector, they pointed the telescope in the direction of VLA 1623. They

FIGURE 5.7 The James Clerk Maxwell Telescope. Credit: R. Phillips/Joint Astronomy Centre

immediately discovered a strong submillimeter source. They knew immediately what this meant. They had discovered an object that is even younger than a Class I protostar, a protostar that is so deeply buried by dust that only submillimeter radiation can escape.

By the time they had driven down to the residence in the morning, of course, all the other astronomers had gone to bed, so they had nobody to tell about their discovery. Nevertheless, they cracked open a bottle of champagne and toasted their discovery as they ate their scrambled eggs. One of them – nobody can remember who – suggested that the object, as it was clearly younger than a Class I protostar, should be called a Class 0 protostar. The paper they eventually wrote describing the discovery of Class 0 protostars is one of the most cited papers in astronomy.

The study of newborn stars, however, got off to a slow start. The main problem was that most telescopes are useless for observing the first moments after the birth of a star – the dust is just too thick. It is only telescopes like the JCMT that operate at long wavelengths that can see through the dust, but even with these telescopes it was still not that easy because the technology at that time (as I write, only twelve years ago) was so primitive.

The instrument that Derek, Philippe and Mary used on the JCMT was the instrument that had been developed at Preston and had been lugged all over the world to different telescopes. With this instrument it was only possible to measure the submillimeter radiation from a single point in the sky. Observers always want to take pictures, but at the JCMT at the time this was just not possible. The only way was to do it indirectly: first measure the submillimeter radiation at one position; move the telescope slightly and measure the radiation at a second position; move the telescope again. . . . and so on, gradually building up a picture of the sky – painting by numbers, as it were.

In 1997 we (by this time, I had been sold on the prospects of submillimeter astronomy) finally got the chance to take pictures, when SCUBA*, the world's first submillimeter camera, was

* SCUBA stands for *Submillimeter Common User Bolometer Array.* You may promptly forget this. Everyone does.

installed on the JCMT. Optical astronomers take cameras for granted because they have been using them for almost 150 years, but submillimeter astronomers have been able to take pictures for less than a decade. A typical optical camera contains an array of detectors (pixels), each of which produces an electrical signal in proportion to the amount of light that falls on it. SCUBA, which was built by a large team of astronomers and engineers in Edinburgh led by the Anglo-Irish astronomer Walter Gear (now also down the corridor from me in Cardiff), worked in basically the same way. It contained an array of detectors that were sensitive to submillimeter radiation. As there were only 37 detectors, it was a pretty basic camera (a typical optical camera contains over one million), but it *was* a camera. Before SCUBA, submillimeter astronomers were in the position that optical astronomers would be in if all their fancy cameras were taken away from them, and instead they were forced to observe the sky with a single pixel or even a single grain of photographic emulsion.

Figure 5.8 shows a picture taken with SCUBA of part of the Orion Molecular Cloud by the Canadian astronomer Doug Johnstone and his collaborators. It is not a conventional picture, of course, because our eyes are not sensitive to submillimeter radiation. The Canadian astronomers have used a computer program to display a visual picture, which our eyes and brains *can* interpret, based on the electrical signals produced by SCUBA – the picture is light where the submillimeter radiation is strong, dark where it is weak. They have chosen to display their image in shades of gray, but they could have used a different by palette of colors and their picture would still have contained the same information about the submillimeter sky. (This is so much the standard procedure in

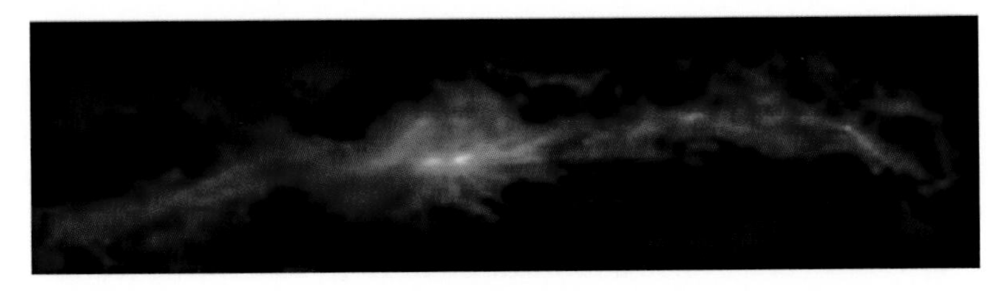

FIGURE 5.8 Image taken with SCUBA of part of the Orion Molecular Cloud. Credit: Johnstone *et al.*

astronomy that even the optical pictures in this book often bear only a tenuous relation to what our eyes would actually see through a telescope.)

The image shows the strange filamentary structure that astronomers also see when they observe the molecular gas. It is possible to see places where there may be protostars, places where the strong submillimeter radiation reveals there is dense dust. This does not imply there is always a Class 0 protostar at the center of the dust. In some places, infrared radiation is also detected, showing the protostar is a later type. At other places, the total amount of radiation summed over all wavebands is not sufficient to indicate a hidden source of energy within the dust. These dense clumps of dust where there is no evidence for a protostar have been given the name *prestellar cores.* As the name suggests, astronomers believe that a prestellar core will eventually collapse to form a star.

SCUBA made it practical for observers to start testing the theorists' models. For example, in the decades before SCUBA, there had been many predictions about how the density of the dust should vary around a protostar. The observers quickly used SCUBA to show that many of these predictions were wrong. The big leap forward, however, was that SCUBA made it possible for astronomers to start the *statistical* study of protostars.

This may not sound terribly exciting, yet statistics has always been vital for making progress in astronomy. SCUBA made it possible for observers to start doing something very simple: count the number of protostars in the different classes. This may still not sound exactly riveting, but let us suppose that the early life of a star does follow the sequence that I have suggested in this chapter: prestellar core; Class 0 protostar; Class I protostar; Class II protostar; main sequence star. The number of objects found at each stage should then reflect the length of time a star typically spends at that stage. Suppose, for example, that stars spend a long time at the Class 0 stage but very little time at the Class I stage. Astronomers should find many more objects in Class 0 than in Class I. Turning this around, if we count the numbers of objects in the different classes, we should be able to learn more about the early lives of stars. (This is what optical astronomers did many years ago when they used the relative numbers of red giants and stars on the main sequence to show a star must spend roughly ten per cent of its life as a red giant.) This census of protostars has not yet proceeded very far, because SCUBA is a rudimentary camera and it takes a long time to survey even a single molecular cloud. However, in 2008 the Herschel Space Observatory will be launched by the European Space Agency. The HSO will overcome the problems faced by ground-based submillimeter astronomers in the simplest possible way by going above the atmosphere. After the launch of the HSO, astronomers will be able to survey every molecular cloud in the sky and so complete the census of protostars in our Galaxy.

In the early 1990s, there was only one large submillimeter telescope. As I write these words in 2006, there are at least ten that have either recently been completed or are under construction. Part of the reason for this efflorescence of submillimeter telescopes was the realization in the 1990s of the huge importance of dust in hiding visible light. As I will explain in Chapter 7, it now appears that roughly half the visible light emitted by all the objects in the Universe since the Big Bang – stars, galaxies and quasars – has been absorbed by dust.

Apart from the HSO, the most important of these new submillimeter telescopes is the Atacama Large Millimeter Array. ALMA will be the ultimate submillimeter telescope in every way: in cost (about half a billion dollars); in location (at 18,000 feet in the Atacama Desert in the Andes); in size and complexity (64 inter-linked telescopes, each about the size of the JCMT – Figure 5.9). By linking the telescopes together to form an interferometer (Chapter 3), astronomers will be able to see vastly more detail in the sky than with a single telescope like the JCMT. To see the gain that ALMA will bring, imagine that you are looking at a painting on the opposite wall of a large hall, so that all you can see is a blur. That is the JCMT view of the submillimeter sky. With ALMA, it will be possible to see every brushstroke in the painting and every grain of paint.

ALMA will make it possible for astronomers to make the connection between the births of stars and planets. Whenever a star is formed, if the standard model (Chapter 1) is correct, a planetary system should also usually be formed. As long as the molecular core is at least slightly rotating, a disk of gas will form around the

FIGURE 5.9 An artist's conception of the Atacama Large Millimeter Array. Credit: Image courtesy of NRAO/AUI and ESO

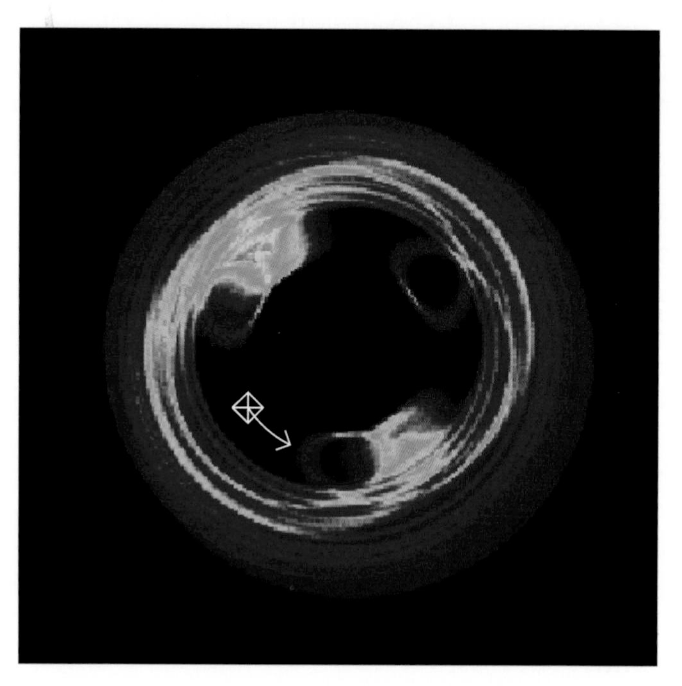

FIGURE 5.10 Simulation of a dusty disk around a newborn star. The clumps in the dust have been produced by the gravitational effect of an unseen planet which is moving inwards towards the star. Credit: Mark Wyatt

protostar, out of which planets will eventually form. The tiny solid particles that initially freeze out of the gas are just dust grains, and so these disks should emit submillimeter radiation. They undoubtedly do so, but with our current blurred view of the submillimeter sky, it is virtually impossible to distinguish the disk around a protostar from the protostar itself.

With ALMA we will finally be able to observe these disks. We will be able to observe them in such detail that we will be able to test many parts of the standard model. It will not be possible to observe planets directly with ALMA, but it will be possible to tell a planet is there by looking for a circular gap in the dust scoured out by the planet. Astronomers will also finally be able to test the idea of planetary migration that I described in Chapter 3. This idea, remember, was proposed to explain why there are so many giant planets close to stars; if the idea is correct, these planets were formed much further out from the star and have since moved inwards. Mark Wyatt at Edinburgh University has shown that one of the results of the gravitational interaction between a migrating planet and a disk will be "clumps" in the dust (Figure 5.10) – clumps that ALMA will be able to detect.

ALMA will open for business in 2012, 25 years after the opening of the JCMT, the world's first submillimeter telescope. In 25 years, submillimeter astronomy will have come a long way.

Part III

Galaxies

6. Silent Movie

Tycho Brahe is a hero of mine because he was the first person to realize the importance of the boring, nitty-gritty little things I do during my day-to-day life as an astronomer: analyzing data, estimating errors, trying to understand the discrepancies between different sets of observations. Other heroes of mine from the dawn of modern astronomy are Copernicus, Kepler, the inventor of the laws of planetary motion, and especially (for me as an observer) Galileo, the first person to look at the sky through a telescope. Nevertheless, although I am in awe of these scientists and fascinated by the stories of their discoveries, I am not very interested in them as individuals. They are too alien from the world of the twenty-first century. They lived too long ago for me to have any understanding or feeling for their personalities – for what made them tick; the times are just too different. The figures from the history of astronomy that *really* fascinate me as individual human personalities are much more modern. These are the scientists who discovered the true scale of the Universe. Because this discovery was made only eighty years ago, these men (and with one important exception they were all men) are only just over the event horizon.

The key discovery was made by an astronomer at the Mount Wilson Observatory in Los Angeles. Mount Wilson is only about twenty miles from Hollywood, and this has always seemed fitting to me because the astronomer, Edwin Hubble, had many of the characteristics of a Hollywood leading man. He definitely had the good looks and the physique. It was said of him once, that if Edwin Hubble had applied to play the part of an astronomer in a movie, he would have been turned down as looking too much like Clark Gable. Unfortunately, he also had some of the Hollywood stars' less endearing traits. He was a terribly vain and egotistical man.

He affected an English accent and tried to airbrush out of his life his perfectly respectable Missouri family. For a time, like all major movie stars, he even had a publicist.

However, if Edwin Hubble was the star in the movie, the movie's producer was George Ellery Hale. Hubble's name is known to almost everyone, if only because of the Hubble Space Telescope, but Hale's name is only widely known among astronomy historians*. Hale however was in many ways a more impressive, and certainly more attractive, figure than Hubble. Many of the institutions of modern astronomy can be traced back to Hale. The International Astronomical Union, the *Astrophysical Journal*, the world's top astronomy journal, and the California Institute of Technology, the world's top science university – all were largely founded by Hale.

Hale's importance in the history of astronomy however was not as a founder of institutions, but as a builder of telescopes. Hale's father was a self-made Chicago businessman, who made his fortune manufacturing the hydraulic elevators necessary for skyscrapers. His father's money and the connections he had to other rich men were a definite help to Hale in his career, but it was not simply money and connections that made Hale the greatest telescope builder of all time. He was also a spectacularly ingenious and resourceful inventor. He designed and built his first laboratory while he was still a teenager, well before his father had made his pile. By the age of twenty two, he had constructed his first astronomical observatory (in his parents' backyard) and had invented the spectroheliograph, still the standard astronomical instrument for observing the Sun. The observatory was of such a high standard that the new University of Chicago immediately made him a full professor, as long as he signed his observatory over to the university. While at the university, Hale founded the Yerkes Observatory outside Chicago and there built a telescope with a lens 40 inches in diameter, making it then the biggest telescope in the world. A large lens makes it possible to collect more light from faint stars (just as a photographer opens a camera's aperture

* This is a different Hale from the one who discovered Comet Hale–Bopp (Chapter 2).

as wide as possible in overcast conditions), and so Hale's construction of the new telescope made Yerkes Observatory the premier observatory in the world – the place to go if an astronomer wanted to peer further out into the Universe.

Hale broke his own record. It is difficult to construct lenses which are bigger than this because a bigger lens will sag under its own weight, but it is possible to build a telescope with a larger aperture by turning from lenses to mirrors. After Hale moved to California, he built a telescope with a mirror 60 inches in diameter on Mount Wilson, then another telescope on Mount Wilson with a 100-inch mirror, and then finally a telescope on Mount Palomar with a 200-inch mirror – each telescope becoming successively the biggest and most sensitive in the world. What is staggering is that almost all this was achieved before Hale was forty. Despite the huge technological projects that he led, he did not have the phlegmatic temperament of an administrator or manager, but instead the highly strung temperament of an artist; in middle age, he had a series of nervous breakdowns which made him, for the rest of his life, a semi-invalid and recluse.

The 100-inch telescope that Hale built on Mount Wilson was the one Hubble used to revolutionize our view of the Universe. A few years ago I visited Mount Wilson. The view from the summit at night was spectacular. In Hubble's time, Los Angeles was a small town and its lights were visible in the distance – Hubble called it the "Los Angeles Nebula." By my time, the plain below the mountain was covered in points and filaments of light. Of the sky itself I could see little, the lights from the city drowning it out. However, even if I could have seen the stars, I doubt they would have been as impressive as the millions of city lights. I remember thinking, as I sat beneath the trees with the dome of the telescope a dark silhouette against the sky and the city spread out beneath me, that even if the number of astronomers who had used the observatory through the years was small, the number of people in the city below who had some connection to the observatory must be quite large. The building of the observatory, in particular the construction of the 100-inch telescope, was one of the big technological projects of the early twentieth century and must have required hundreds of technicians and engineers. And running an observatory does not just require astronomers, but also technicians, cleaners, telescope operators, janitors, and even people to cook meals for the rest of the staff.

The following day I drove up to the observatory again. This time the most spectacular view was of the smog blanketing Los Angeles; the smog and the bright city lights make the observatory only useful now for some specialized types of research. I went into the dome of the 100-inch. Like all telescope domes during the day, there was something of the feel of an abandoned movie set: nobody around; lots of bits of discarded equipment of no apparent purpose. There was a smell of machinery and a fusty feel to the place. The telescope itself looked like it had been constructed from a 1920s Meccano set. I couldn't imagine that anyone had ever really used it.

However, despite the air of a faded movie set, the 100-inch *was* the technological marvel of its age, and building it led to one of Hale's nervous breakdowns. Among many technical challenges, the big one was the mirror: casting such a large disk of glass and grinding it into precisely the right shape. The disk of glass out of which the mirror was ground was initially rejected by Hale because it contained a large number of small air bubbles, and it was only after a second disk cracked in two that the first was re-examined and it was decided that it might be suitable after all. The grinding and polishing of the disk took five years. The moment of truth for any telescope is the moment that it is turned on the sky for the first time. Until that moment it is always possible, especially since telescopes are always built on the technological edge, that it may not work at all – with the loss of millions of dollars, reputations, and careers. One of the more embarrassing moments for modern astronomers was when the Hubble Space Telescope was turned on a star for the first time and it was discovered that the mirror, because of an incorrect sign in a computer program, had been ground into the wrong shape (nobody had checked the shape of the mirror until the telescope was in orbit!). For the 100-inch, this moment of truth was on November 1st, 1917.

There were nineteen people present in the dome of the 100-inch that night, ranging from Hale to the observatory's janitor, Roy Desmond, and including the English poet, Alfred Noyes, whom Hale had invited along in the hope that he might write a poem about the occasion*. Hale and the assistant director, Walter S.

Adams, climbed to the observing platform. They decided to look first at one of the brightest objects in the sky, the planet Jupiter. Hale called out a command to the telescope operator and the barrel of the telescope slewed towards the planet. When the telescope came to a stop, Hale bent low and looked through the eyepiece. He stood up without saying a word. Adams bent down to have a look. Instead of a single image of Jupiter, he saw six blurred images. An obvious possibility was that the mirror, perhaps because of the air bubbles in the glass, was fatally flawed. It is impossible to know what thoughts passed through their heads as they tried to understand the reasons for this disaster, but they must have included the possible consequences: the drying up of funds, ruined careers. Eventually, however, someone suggested a second, less catastrophic possibility. During the day the dome had been left open while workmen had worked on the telescope, and it seemed just possible that the Sun had shone on the mirror cover, heating up the mirror and distorting its shape. The only way to see whether this was true was to wait until the cold night air had cooled down the mirror. However, after they had waited in the cold dome for several hours in increasing anxiety and boredom, there had been little improvement in the images. Eventually they all went to bed. Hale and Adams agreed to meet back in the dome at 3 a.m. Hale lay down without undressing. After an hour of tossing and turning, he turned on the light and tried to read a detective story. But this failed to stop his racing mind, and he eventually gave up and went back to the dome. Adams was already there; he hadn't been able to sleep either. Jupiter had now set, so they pointed the telescope at the bright star Vega. Hale looked through the eyepiece again. This time Vega appeared as a single sharp point of light.

Hale's dream of building bigger and bigger telescopes was not based on satisfying his own ego, although in such a complex personality this may have been a part. At the beginning of the twentieth century, the increased sensitivity provided by telescopes with big mirrors was needed to answer several important scientific questions. The grandest of these questions was the mystery of the nebulae.

* He did. It was not one of Hale's better ideas.

The view of the sky from Mount Wilson today is similar to the view that most of us have. Today most people in industrialized countries live in cities, and even the lights of a village are often enough to drown out most of the six thousand stars that would be visible deep in the country. Nebulae are particularly difficult to see from within a city; the only nebula that is visible from my back garden is the Orion Nebula (Chapter 5). The other thing that is missing from my view is the Milky Way, the faint band of light that stretches across the sky, which can only properly be seen away from the city lights. Using one of the first telescopes, Galileo showed that the Milky Way is the combined light from countless faint stars and not, as the Greeks thought, a jug of milk spilled across the sky by a god. One hundred and fifty years later, Thomas Wright suggested that the Sun is one of millions of stars in a huge disk, and that the Milky Way is a trick of perspective: if you look towards the Milky Way, you are looking along the plane of the disk and see the combined light of millions of stars; if you turn your head ninety degrees, you are looking out of the disk and see comparatively few stars. This huge agglomeration of stars, our home in the Universe, is the *Galaxy**.

The Orion Nebula and many other nebulae are close to the Milky Way and are clearly clouds of gas within the Galaxy. But there are other nebulae which are not so obviously within the Galaxy. In the middle of the nineteenth century, astronomers discovered nebulae which looked quite different from the Orion Nebula and the other Galactic gas clouds. These nebulae did not look like shapeless clouds of gas but instead had a distinctive spiral structure. They were also mostly found away from the Milky Way. As far back as the mid-eighteenth century, the philosopher Immanuel Kant had suggested that some nebulae are galaxies like our own but are so far away that their individual stars can not be seen – with the faint fuzz of light thus being the combined light of billions of stars. By the end of the nineteenth century there were tens of thousands of nebulae known, and astronomers were divided into two camps: those who believed all the nebulae are

* From *gala*, the Greek for milk.

within the Galaxy and those who believed that some of the nebulae, especially the spiral nebulae, are galaxies themselves.

Just before the First World War, a spectacular discovery was made about the spiral nebulae. In 1912, an astronomer with the splendid name of Vesto Melvin Slipher* began to measure the spectra of the spiral nebulae. The nebulae are very faint, so obtaining their spectra required very long exposures, but eventually Slipher managed to obtain spectra for forty. He was then able to measure their speeds by looking for changes in the wavelengths of the spectral lines – the Doppler shift, the same technique that modern astronomers have recently used to show there are planets around other stars (Chapter 3). He announced his results in 1914 at the annual meeting of the American Astronomical Society, which was in Chicago and attended by Edwin Hubble, about to start a Ph.D. at nearby Yerkes Observatory. To everyone's surprise, of the forty nebulae, almost all had "redshifts," with the spectral lines shifted to the longer wavelength, red end of the spectrum. Slipher's results showed that the nebulae are moving away from the Earth at tremendous speeds. Everyone realized this discovery was important (Slipher received a standing ovation, something I have definitely never seen at an astronomy meeting), but nobody knew quite what to make of it. The high speeds strongly suggested that the nebulae are not part of the Galaxy. But if they are separate galaxies, why are they all moving away from us so quickly?

The person who provided the key to solving the mystery of the nebulae traditionally only has a small part in this story. For the movie version of the *Mystery of the Nebulae*, I envisage the following scene. The scene is a dark and dusty back room at

* Other good names are Solon P. Bailey and, unfortunately only an amateur, Amor Cosmo. The latter started out life as plain Bill Smith. Because of the difficulty he had in receiving the correct mail in Californian mining camps – Bill Smiths are too common – and because of his interest in astronomy, he changed his name to *Amor Cosmo*, Latin for "lover of the universe". He later became premier of British Columbia, showing that politicians don't have to have gray personalities.

Harvard College Observatory. There is a woman in the room stooped over a photographic plate staring intently at it through a microscope. She is a spinster in her late thirties with her hair scraped back in a bun and wearing a white dress. There is a knock at the door. The director of the observatory, Edward Pickering, an elderly man with a large white beard, comes in. He addresses the woman formally as Miss Leavitt and gives her fresh orders and a new batch of plates. This scene would provide some valuable early twentieth century atmosphere for the movie, but since there is very little action, after a minute or two at the most I would cut away to a scene showing the star actors on Mount Wilson.

The bit-part that I have given Henrietta Leavitt in the movie, and that she has in the history books, is unjust. She was one of Pickering's "computers" – women with a college education who received pitiful wages for doing calculations and other menial work for the male astronomers. Her job was to analyze photographic plates taken with Harvard's telescope in Peru. Her years sitting in that back room, carefully measuring the sizes of stellar images on photographic plates (*sitzfleisch*, a German word meaning the ability to keep your bottom planted on a chair, is one of an astronomer's most useful qualities), led to one of the most important discoveries in the history of astronomy. However, because she was merely a computer, she is little more than a footnote in scientific history. The wider scientific community only heard about her work through Edward Pickering; she wrote no autobiography and left no letters; there is very little known about her. She is one of the many people in history – proportionally more of them poor and female – without a voice*.

The fundamental discovery that Leavitt made, in the early years of the twentieth century, was about a type of star called a *Cepheid variable*. In the eighteenth century, an amateur astronomer named John Goodricke noticed that a star in the constellation Cepheus, δ Cephei, is a variable star; its brightness increases quickly and then slowly falls, repeating the cycle every few days. Cepheid variables are stars that vary in the same

* The little that is known about Henrietta Leavitt is described by George Johnson in *Miss Leavitt's Stars*.

characteristic way as δ Cephei, although the cycles of different stars vary in length between one day and several weeks. The usual metaphor for any periodic variation is a beating heart, which is actually highly appropriate in this case because a Cepheid's variation is caused by the whole star expanding and contracting. Leavitt found many Cepheid variables on the plates she was studying. As a diligent computer, she naturally measured the period of variation of each star (the time between two common points in the star's cycle). The brightness of each star obviously varied, but she was still able to measure an *average* brightness for each star. The thing that raised her from the ranks of the many careful and diligent observers throughout history is what she did next.

To understand this, it is important to recognize the distinction between brightness and luminosity. The luminosity of a star is essentially the total amount of light it radiates; its brightness is how much light is received from the star by us on the Earth. Thus several stars may have the same luminosity, but because they are at different distances from the Earth, have a very different brightness. The Sun, as it happens, is not a very luminous star, but because it is so close it is a very bright one.

It is easy enough to measure the brightness of a star, but to find its luminosity (for a variable star like a Cepheid, the *average* luminosity) it is also necessary to know its distance. Leavitt did know the distances of some Cepheids, and so she was able to calculate their luminosities. When she had assembled a reasonable sample of Cepheids, she plotted a graph of the luminosities of the stars against their periods (Figure 6.1). She discovered the interesting fact that there appears to be a relationship between the two: the Cepheids with higher luminosities have longer periods. A Cepheid, for example, that has a luminosity 4000 times greater than the Sun (Cepheids are very luminous stars) has a period of about 12 days, but one that has a luminosity only 600 times greater than the Sun has a period of only four days.

This is an interesting fact, but the reason the relationship is so important – the reason why Leavitt's name is at least mentioned in the history books – is that it immediately gave astronomers a way of measuring distance. Suppose you want to measure the distance of a faint star cluster. After you have taken several images of the cluster, if you are lucky you will find a star that varies in

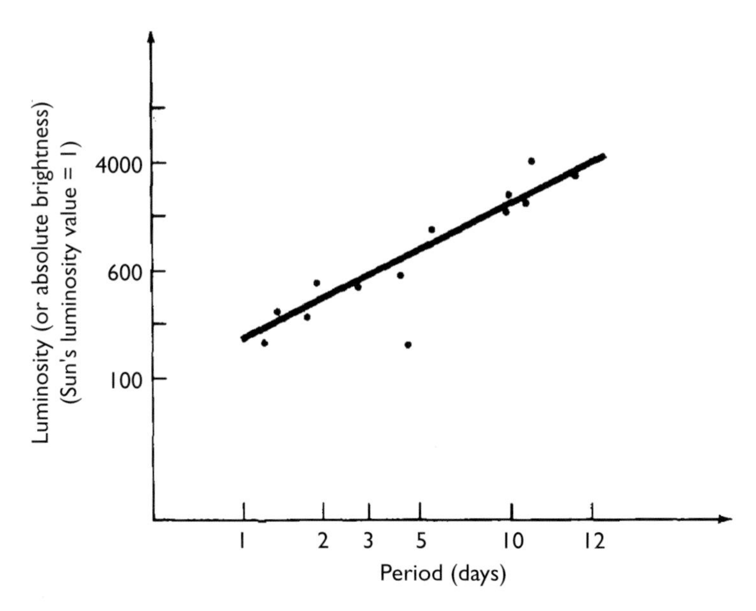

FIGURE 6.1 The luminosity of a Cepheid variable star verses its period of variation. The more luminous stars have longer periods. Credit: Eric Chaisson and Prentice Hall

the characteristic way of a Cepheid. Once you have found your Cepheid the rest is easy. Measure the period of the Cepheid and its average brightness. From Leavitt's graph and your measurement of the period, estimate the luminosity of the star. Once you know the luminosity and brightness of the star, it is then a simple matter to calculate its distance, which also of course gives you the distance of the cluster. The method is beautiful in its simplicity, but in practice it requires a very sensitive telescope to find the Cepheids in a distant cluster.

Leavitt had a bit-part in the *Mystery of the Nebulae.* In 1915 one of the stars in the movie walked onto the set at Mount Wilson. He was from Missouri; he revolutionized our ideas about the Universe; he was very vain. Curiously, he was not Edwin Hubble.

If Hubble resembled a movie star from the early days of Hollywood, Harlow Shapley was a much more modern kind of movie star. With his matinee-idol good looks and fake English accent, Hubble had the slightly unreal persona of the early movie stars. Shapley, who never lost his Missouri accent and had once worked as a crime reporter, was a more naturalistic kind of actor – a tubby Robert Mitchum perhaps rather than Douglas Fairbanks.

He was born in 1885 in an America in which the acceleration pedal was hard down. In the west, there was still the Wild West (the Battle of Little Big Horn had only been nine years before). In the east, America was changing from a country in which almost everyone was born on the family farm to one of industries and cities. It was a country of fabulous contrasts. Immigrants from Europe flooded into the cities and often ended up living ten in a room in tenement blocks; industrialists like John D. Rockerfeller and Andrew Carnegie lived like the more decadent Roman emperors. It was a country in which money talked*, in which skyscrapers were reaching for the clouds, but also one in which it was still possible to be scalped. In this rapidly changing America, in which many of the famous universities of today had not yet been founded, there were few clear career paths. Ambition, energy, intelligence and a slice of luck would usually take you somewhere, but it was not always where you intended.

Shapley himself was born on the family farm. He left school at 15 and became a crime reporter in a nearby town in the middle of an oil boom. It was here that he saw a public library for the first time, and much of his education was obtained sitting in this library, soaking up history, literature, and poetry. After a time, he decided he needed some more formal education and he enrolled at the University of Missouri where he intended to study journalism. When he arrived at the university, he discovered the opening of the new school of journalism had been postponed for a year. Forced to choose other courses, he looked through the alphabetical course listing, coming first to archaeology, a word he couldn't pronounce. Deciding that if he could not pronounce the name of a course he shouldn't take it, he looked at the next course in the list, which was astronomy. This sounded interesting and he could pronounce the word. He enrolled in the course and never looked back.

He arrived on Mount Wilson at a time of opportunity. Leavitt's discovery had provided a method for measuring distance, and the new big telescopes on Mount Wilson made it possible to apply this method to faint star clusters. Shapley decided to tackle

* "The Pennsylvania legislature is the finest body of men that money can buy."

a subject about which little new had been learned in two centuries: the structure of the Galaxy.

In the two centuries since Thomas Wright's insight that the existence of the Milky Way implies we live in a disk of stars, only one new thing had been learned about the Galaxy. Several astronomers, in particular the Dutch astronomer Jacobus Kapteyn, had used telescopes to count the number of stars in different parts of the Milky Way. They always found roughly the same number of stars no matter which part of the Milky Way they looked at. They concluded that this must mean we live close to the center of the Galaxy, because otherwise we would see more stars in one direction than another.

Shapley's big plan for surveying the Galaxy was based on measuring the distances of globular clusters. There is no globular cluster visible to the naked eye, which is a pity because a globular cluster is one of the most beautiful things in the sky; viewed through a telescope, a globular cluster, a spherical cloud of up to a million stars, looks like a cluster of tiny jewels. After arriving on Mount Wilson, Shapley spent several years using the 60-inch and 100-inch telescopes first to find the Cepheid variables in globular clusters and then, using Henrietta Leavitt's measuring rod, to measure the distances of the clusters. By the end of this time, he had measured the distances of most of the hundred-or-so globular clusters that were then known. He discovered that the globular clusters were themselves distributed in a spherical cloud. This cloud however was not centered on the Sun, but on a point about 60,000 light years away. Shapley's intuition, which has since been confirmed in many ways, was that this is the true center of the Galaxy.

Shapley's discovery pushed humanity away, once again, from a special place in the Universe. Copernicus had showed that the Earth is not at the center of the Universe, but merely one of several planets orbiting the Sun; Shapley's discovery showed that the Sun is not even close to the center of the Galaxy but merely a humdrum star somewhere in the Galactic suburbs. The reason Kapteyn and others were wrong is interstellar dust, which absorbs starlight (Chapter 5). If it were not for dust, the center of the Galaxy would be obvious. The Milky Way would be brightest in the constellation Sagittarius, which is the direction of the Galactic center, but the dust hides most of the stars in this direction.

Shapley's discovery, which he announced in 1918, was one of the greatest astronomical discoveries of the twentieth century.

The following year a second movie star arrived on Mount Wilson.

One summer morning in 1919 Donald Shane, a Ph.D. student in astronomy, happened to be waiting at the side of the street in San Jose, California, for the coach to take him to Lick Observatory on Mount Hamilton. Across the street he noticed a tall handsome man in military uniform. The man came over and introduced himself. Shane recognized the name from scientific papers from the Yerkes Observatory, but he was confused by the British accent. The man explained that he was on his way to join the staff at Mount Wilson but had decided to visit Lick Observatory along the way. Shane remembered that Edwin Hubble spent only a day on Mount Hamilton, but it was enough to make a lasting impression on the astronomers there. For the rest of their lives, the astronomers who were at Lick Observatory that day would refer to Hubble as "the Major."

Hubble arrived on Mount Wilson in 1919, Shapley left in 1923. They spent four years together at the observatory. There is however almost no record of what they thought of each other. They should have despised each other because they were polar opposites in almost every way. They *were* both from Missouri, but unlike Hubble who wouldn't even have his relatives in the house, Shapley was proud of his poor background and had kept his accent. Shapley had a sense of humour, Hubble did not; Shapley was a pacifist, Hubble fought in the war; Shapley was a liberal democrat, Hubble was a right-wing republican. The only thing they seem to have shared was vanity. But, apart from a remark by Shapley that Hubble "was a Rhodes scholar and didn't live it down," there is nothing on record about either man's personal feelings. The prosaic reason for this is probably that this was a tight-lipped era, and anyway scientists are supposed just to look at the *work* and ignore each others personal foibles. But I prefer to explain this in terms of the movies. Hubble and Shapley were such different types of movie star that they could never have been in the same movie. For the four years during which they were both on Mount Wilson, Shapley was the star of the movie. Hubble was somewhere else on the film lot, waiting in his trailer.

In astronomical circles, Shapley's name is still a famous one. Nevertheless, he never became the household name that Hubble later became. The reason for this is that Shapley made three mistakes. If not for these, it would probably now be the Shapley Space Telescope orbiting the Earth. Shapley's first great mistake was to take on the role of prosecuting lawyer in one of the most famous trials in astronomy.

Shapley's discovery of the huge size of the Galaxy (300,000 light years from end to end) had made him a convert to the idea that the spiral nebulae could not possibly be separate galaxies but must lie within the Galaxy. When in 1920 George Ellery Hale arranged a debate on the nature of the nebulae at the National Academy for the Advancement of Science in Washington, Shapley agreed to put the case that the nebulae lie within the Galaxy – with Heber Curtis from Lick Observatory putting the case that the nebulae are separate galaxies. The "Great Debate" has entered the mythology of astronomy, and with the hindsight of eighty years it is tempting to think that Shapley must have been stupid to have chosen the wrong side of the debate. But after vaguely knowing about the debate for twenty years, when I finally read the details, I realized that the intellectual arguments on Shapley's side were quite strong, and that he was unlucky that he went down in history as the brilliant trial lawyer on the wrong side of the case.

The main piece of evidence in Shapley's case was the testimony of another Mount Wilson astronomer, Adriaan van Maanen. By comparing photographs taken at different times of the same spiral nebula, Van Maanen had discovered that the bright knots in the nebula were moving, and he concluded that the nebula was rotating. However, if the nebula is outside the Galaxy, such movements would be much too small to detect. Another piece of evidence was a nova that had appeared in the Andromeda Nebula in 1885, which for a while had been as bright as the entire nebula. A nova is an explosion in which a thin layer of gas on the surface of a star is thrown into space. Shapley argued that a nova could not possibly be as bright as an entire galaxy. He was very unlucky. Both of his arguments turned out to be wrong, but for reasons that were not his fault. It gradually became clear that Van Maanen's testimony was invalid; he had not been careful enough in his measurements of the nebula. Shapley's other argument was completely correct *if* the star that had appeared in Andromeda in 1885 had been a nova, but it was actually a supernova, an explosion in which an entire star is ripped apart (Chapter 4). Unfortunately for Shapley's case, a supernova *can* outshine an entire galaxy.*

Shapley's second great mistake is more puzzling. Shapley had made his name using Miss Leavitt's measuring rod to measure the distances of globular clusters. It would have been possible to do the same for the nebulae themselves, which would immediately have answered the question of whether the nebulae are within the Galaxy. Shapley never did this. There is also a story that makes his failure to do this seem even stranger.

It has to be admitted that the source of this story is not the most reliable. Milton Humason was another Mount Wilson astronomer with an interesting resumé. He was one of the thousands of people who built the observatory, working as a mule driver on the pack trains carrying construction material up to the summit. Falling in love with life on the mountain (and also with the daughter of the observatory's engineer), he got a job as the observatory's janitor. He soon showed such great ability at taking astronomical photographs and at enduring the cold of the Mount Wilson nights that he got hired as an observer. A hunter, a card-player and a drinker, Humason told this story about Shapley's last days on Mount Wilson.

Shortly before Shapley left Mount Wilson he gave Humason some photographic plates of the Andromeda Nebula to compare on the stereocomparator. The purpose of the stereocomparator, the device Clyde Tombaugh used to find Pluto (Chapter 2), is to make it easy for the eye to pick out objects on a plate that are either moving or varying. Humason loaded the plates and immediately found several stars he was sure were Cepheid variables – if so, they could be used to measure the distance of the nebula. He marked

* Curtis' evidence seems to me less convincing, although of course he was right. His main argument was that the spectra of the spiral nebulae resemble those of stars rather than clouds of gas. Curiously, neither scientist seems to have made much of the redshifts found by Slipher, probably because, whether the nebulae are inside or outside the Galaxy, nobody could think of a way of explaining them.

the positions of the stars in ink on the plates and excitedly went off to tell Shapley. He claimed that Shapley then simply repeated his arguments from the Great Debate that the nebulae must be within the Galaxy. According to Humason, Shapley then took out his handkerchief, turned the plates over, and wiped them clean of Humason's marks.

I do not believe this story. For it to be true, Shapley would have to have been crazy, and neither his career before or after the event gives any reason to think he would have been guilty of such bizarre behavior. I think Humason's story was just a tall story dreamt up because nobody could quite understand why Shapley had not taken the small step of extending the Cepheid distance technique, which he had pioneered, from globular clusters to the nebulae. My suspicion is that the true explanation is the gentlemanly traditions of early-twentieth century science (there were no women, of course, apart from humble computers like Henrietta Leavitt). In this still rather Victorian world, it was regarded as "bad form" if you infringed on somebody else's research field. Hubble had done his doctoral thesis on the nebulae using the 20-inch reflector at Yerkes Observatory, and he had been taken on to the staff at Mount Wilson on the strength of this thesis. If Shapley had extended his distance technique to the nebulae, he would have been trespassing on Hubble's territory. To my mind, as a modern astronomer, this seems ridiculously scrupulous, and I suspect it was even by the standards of the early twentieth century. But Shapley was still, underneath, the poor boy from the farm and, despite his radical politics, he would probably have been less likely to infringe the customs of this upper-class world than somebody who had grown up with them.

Finally there was Shapley's third great mistake. In 1923 he was offered the top job in astronomy: the directorship of Harvard College Observatory. To the boy with the country accent who had not known how to pronounce the word "archaeology," this job at the most prestigious university in the country must have been irresistible. However, Shapley's mistake was that this was not the top job any longer. In the modern era in astronomy the crucial thing for an observer is access to big telescopes. At Harvard, Shapley had the prestige but no large telescopes. Hubble never made this mistake. In years to come, when he was unpopular with his colleagues at Mount Wilson, when he was passed over for the directorship of the observatory, when he was offered much larger salaries at universities on the East Coast, he stayed with the big telescopes on the mountain.

Exit Harlow Shapley stage left, enter Edwin Hubble stage right.

On October 4th, 1923, Edwin Hubble was in the dome of the 100-inch telescope on the penultimate night of an observing run. The weather was poor, but not bad enough to close the telescope and go to bed. Hubble asked the night assistant to point the telescope at one of the spiral arms of the Andromeda Nebula. When the telescope reached the correct position, Hubble loaded a new photographic plate into the telescope's plate holder and started a forty-minute exposure. After the exposure had finished, he started a new one and took the exposed plate down to the telescope's dark room. At night, a telescope's dark room is a lonely place; there are no windows and the only other person around, the night assistant, is usually several floors away. Hubble would have spent close to an hour in the dark room, passing the plate through the series of chemical baths necessary to turn it into an image of the sky. At the end of this process, he would have seen an image of the Andromeda Nebula, one of the most beautiful of the spiral nebulae – a beautiful image, but one he would have seen many times before.

This time however he saw something different. There was a bright star in the nebula he had not seen previously. He suspected it was a nova. This was interesting but nothing exceptional. Novae had been seen in nebulae before, and indeed a previous nova in Andromeda had been one of Shapley's arguments for why the nebula could not be too far away.

On the following night, the weather was much better. Hubble took a slightly longer exposure of the same part of the Andromeda Nebula he had observed on the previous night. When he developed the plate he found two additional new bright stars, which again he suspected were novae. This was the last night of his observing run, and at the end of it he drove down the mountain and went to bed.

After sleeping, he went to his office and looked through the observatory archive for previous plates of the nebula. When he compared these earlier plates with his new ones, he discovered that one of his three suspected novae could not be a nova at all because it was there on the previous plates. It was however clearly varying in brightness. From the way that its brightness was varying – a rapid increase and then a gradual fall, with the cycle repeated every few days – he realized it must be a Cepheid variable. From the images on the different plates, he estimated the period and average brightness of the star. And then, sitting in his office on October 6th, 1923, Hubble used Henrietta Leavitt's period–luminosity graph to measure the distance of the Andromeda Nebula. The value he obtained was one million light years. Since the size of the Galaxy, according to Shapley, was three hundred thousand light years, Hubble's measurement showed that the nebula must lie well outside the Galaxy. The Andromeda Nebula is a galaxy in its own right.

The Andromeda Nebula, which should now be called the *Andromeda Galaxy*, is actually remarkably like our own. Both are spiral galaxies of a similar size. They are the largest galaxies in a group of about forty galaxies which is called by astronomers the *Local Group*. The smaller galaxies in this group swim around the two large galaxies like fish around two whales.

Hubble's discovery expanded the human horizon however far beyond this nearby group of galaxies. Andromeda is the brightest galaxy in the northern sky and, outside a city, can just be seen with the naked eye. The faintest nebulae visible on Hubble's plates were about one million times fainter than this. It is impossible to find Cepheids in nebulae this faint, but Hubble was able to estimate their distances by assuming they are galaxies similar to Andromeda. The brightness of an object scales with the inverse square of its distance: if the object is moved twice as far away, its brightness falls by a factor of four; if it is moved three times as far away, its brightness decreases by a factor of nine, and so on. Because the faintest nebulae were one million (a thousand thousands) times fainter than Andromeda, Hubble realized they must be about one thousand times further away. Using his measurement of the distance of Andromeda, he calculated that the faintest nebulae on his plates must be at a distance of about one billion*

* One billion is one thousand millions.

light years. From the number of faint nebulae he could see on his plates, he estimated that the Universe must contain at least one hundred million galaxies. Thus one day in 1923, in a nondescript Los Angeles office, the human horizon exploded outwards: from a single galaxy to a Universe of at least one hundred million galaxies, one billion light years in size.

Hubble's discovery made him a celebrity. He won the annual prize given by the American Association for the Advancement of Science, but he was also invited to dinner by Charlie Chaplin. He appeared in the New York Times and Life Magazine, and there was a constant flow of movie stars from Hollywood up to Mount Wilson to have their pictures taken with the famous astronomer. Hubble and his wife were even invited to spend weekends at the San Simeon estate of William Randolph Hearst, the flamboyant publisher and the model for Citizen Kane. An invitation to San Simeon, roamed by herds of Zebra and Impala, where the guests were waited on by footmen and ate caviar off blocks of ice, was a sure sign that one had reached the celebrity A-list.

Shapley heard about Hubble's discovery in a letter from Hubble, one which I imagine Hubble must have enjoyed writing. A student happened to be in Shapley's office when he received the letter. After reading it, Shapley passed the letter to her, saying, "Here is the letter that has destroyed my universe."

The remainder of Hubble's life was spent exploring this new enlarged Universe. His expedition through this Universe was a surprisingly solitary affair. Today, there are probably about one hundred large optical telescopes used by several thousand professional astronomers. In Hubble's time, there were very few telescopes large enough to be useful for observing galaxies, and until the late thirties most of the exploration of this new Universe was done by Hubble himself.

It is impossible here to resist poetry. The image of Hubble that always comes into my mind is of him alone in the dark room of the 100-inch late at night, crouched over plates showing thousands of faint galaxies. This always makes me think of Columbus entering the New World. Hubble himself thought like this, describing his work as a "dream" and an "adventure," and calling his book on his discoveries *The Realm of the Nebulae.*

The maps that Hubble made encourage this fancy. The Galaxy is mostly empty space. The space between the stars is much greater than the sizes of the stars; if the Sun were a football, the nearest star would be another football at about the distance of Hawaii (Chapter 4). Hubble however found that the space between two neighboring galaxies is generally only about one hundred times greater than the sizes of the galaxies themselves, evoking an image of galaxies as an archipelago of Caribbean islands, each within easy sailing distance of the next.

In this New World, Hubble discovered a variety of inhabitants. Galaxies are members of three main tribes. The first tribe are the spiral galaxies; the second are galaxies which appear round or oval on photographic plates – using his right as the discoverer, Hubble named these *ellipticals*; the third tribe are galaxies with no regular structure at all – Hubble named these *irregulars*.

He discovered that the members of the same tribe are often very different. The spiral tribe consists of two separate clans: the *barred spirals*, which have bars of light across their centers, and the spirals without bars. Members of both clans have central round or oval assemblies of stars – Hubble named these *bulges*. He discovered that in some spirals the bulges emit more light than the surrounding spiral arms, whereas in others the bulges are tiny and hardly visible. In some spirals the spiral arms are tightly wound around the bulge; in others the spiral arms look more like flowing tendrils. Figure 6.2 shows some pictures of the spiral tribe taken from the book that Hubble wrote about his discoveries in this New World.

He also discovered that members of the same tribe have very different sizes. A dwarf elliptical galaxy contains only about a million stars, making it hardly bigger than a globular cluster; a giant elliptical galaxy however may contain as many as one thousand billion stars. He discovered that the tribes congregate in different places. Most galaxies are found in small family groups like the Local Group, but some galaxies, ellipticals in particular, are members of clusters that contain hundreds of members (Figure 6.3).

Only six years after proving the existence of galaxies, Hubble made another monumental discovery.

FIGURE 6.2 Some of the pictures of spiral galaxies Hubble showed in his book *The Realm of the Nebulae*. The three galaxies on the right are barred spirals, although the galaxy at the top right has such a large bulge that the spiral arms and bar are barely visible. The galaxy at the top left is a spiral galaxy seen from the side. The spiral arms are not visible and the dark band across the galaxy's bulge is caused by interstellar dust, which absorbs visible light (Chapter 5).

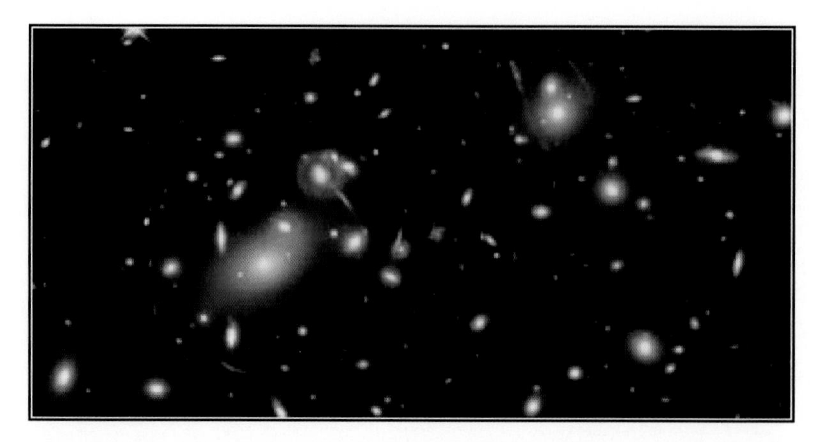

FIGURE 6.3 A picture of a cluster of galaxies, taken this time not by Hubble but by the Hubble Space Telescope. There are several hundred galaxies in this cluster. Almost all of them are ellipticals. Credit: W. Couch, R. Ellis and NASA

The story of the redshifts is a labyrinthine one. As I described earlier, in 1914 Vesto Slipher showed that most of the nebulae have redshifts and are therefore moving away from us. Two years later, Albert Einstein published the general theory of relativity. I will describe this theory in a little more detail later in this chapter, but for now all that the reader needs to know is that the general theory of relativity is, like Newton's theory, a theory of gravity.

One of the first things Einstein did after inventing this new theory of gravity was to apply it to the Universe as a whole. To his surprise, he found that the effect of gravity is that the Universe can never be at rest. Suppose the Universe *is* momentarily at rest and none of the galaxies in it are moving. The gravitational attraction between the galaxies will inevitably make them start moving towards each other – and so the Universe will begin to contract. The Universe may be expanding, although gravity will work to slow the expansion, or it may be contracting. But it can never stand still for more than an instant.

Einstein however did not believe any of this. The permanence of the night sky, apart from very rare events such as supernovae, suggests that the Universe is not expanding or contracting. Since Einstein *knew* that the Universe is eternal and at rest, he did not believe the prediction of his own equations. For no good reason, apart from the need to balance the force of gravity and allow the

Universe to stand still, he introduced an extra term into the equations. He later called the introduction of this extra term, the *cosmological constant,* his "greatest blunder"*.

The startling thing is that this was now several years after Slipher had discovered that most of the nebulae are moving away from us. At the time, many astronomers believed the nebulae were galaxies (although this was still before Hubble had proved it), and it is remarkable that it was not immediately realized that the redshifts of the nebulae and Einstein's theory were pointing to the same thing: an expanding Universe. What is even more surprising is the length of time it took for everything to become clear: not until 1930, fourteen years after the publication of the theory of relativity and sixteen years after Slipher's discovery of the redshifts.

The next character in the story is a tragic figure: Alexander Friedmann, a Russian mathematician who died young. In 1922 Friedmann solved Einstein's equations correctly, without making the mistake of introducing a cosmological constant, and he discovered that the Universe is dynamic – it must either be expanding or contracting. However, he published his work in an obscure Russian astronomical journal read by almost nobody outside Russia. Five years later a Belgian priest, George Lemaitre, who had not heard of Friedmann's work, repeated his calculations. But he too published his work in an obscure journal, and again very few people learned about it. Friedmann and Lemaitre were not part of the scientific establishment; the director of Leiden Observatory in the Netherlands, Willem de Sitter, was definitely a member of the international scientific club. He too had solved the equations of general relativity and developed a model for the Universe. De Sitter's model was less realistic than the models of Friedmann and Lemaitre because it contained no matter, but it made the inter-

---

* Einstein's blunders were not of course like those of ordinary scientists. Over the last eighty years, the cosmological constant has been repeatedly rescued from the rubbish bin of history as a way of explaining discrepant observations. It is now once again in fashion and, as I will describe in Chapter 8, it now looks as if Einstein was wrong about the Universe being at rest, but right about there being something like a cosmological constant.

esting prediction that the redshifts of the galaxies should depend on their distances (as did the unknown models of Friedmann and Lemaitre). In the summer of 1928, Edwin Hubble, by now another heavyweight member of the international scientific establishment, visited de Sitter in Leiden and was persuaded by him to test this prediction.

When Hubble returned to Mount Wilson, he set about testing de Sitter's prediction. This was much more difficult than showing the nebulae were galaxies – for this he had only needed to measure the distance of a single nebula. To test de Sitter's prediction, Hubble needed to measure the distances of many galaxies, including quite distant ones – and this was beginning to push the capabilities of the 100-inch telescope. Nevertheless, Hubble had already found Cepheids in several other galaxies apart from Andromeda, and he was soon able to show that the speed and distance of a galaxy are related: the more distant galaxies are moving faster away from us (Figure 6.4). In mathematical terms, the speed of a galaxy is equal to its distance multiplied by a constant, which in honour of Hubble is now called the *Hubble constant*. About the

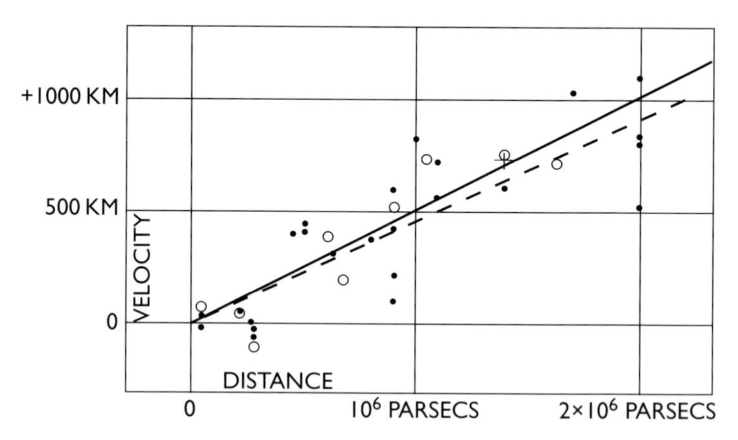

FIGURE 6.4 Hubble's diagram, taken from *The Realm of the Nebulae*, showing the relationship between the speed of a galaxy and its distance from us. The speed of the galaxy in kilometers per second is plotted along the vertical axis. The distance of the galaxy, measured in the astronomical unit of parsecs (one parsec is about three light years), is plotted along the horizontal axis. The most distant galaxies in the diagram are about six million light years from the Earth and are moving away from us at about 1000 kilometers per second.

same time that Hubble was doing this, Lemaitre met the great physicist Arthur Eddington at a conference in London and managed to interest him in his work. Eddington then wrote a popular article about Lemaitre's work, which, to round the story off, is how Hubble heard about it. The combination of Slipher's discovery of redshifts, Hubble's discovery of the relation between speed and distance, and the theoretical work of Lemaitre finally convinced the scientific community that we live in an expanding Universe.

The importance of Hubble's discovery of the relation between speed and distance, now called "Hubble's law," is that it shows our Galaxy is not a special one. At first sight, the fact that almost all galaxies have redshifts seems to suggest just the opposite: all the other galaxies are moving away from ours, and so ours must be unique. Let us suppose, however, that the Universe is uniformly filled with galaxies. The upper part of Figure 6.5 shows a small region of the Universe containing a handful of galaxies; the lower part of the figure shows the same region at a later time after the Universe has expanded in size by a factor of two. The distance between any pair of galaxies in the figure will also have increased by a factor of two. Now choose a galaxy to live in. Whichever galaxy you have chosen, you will find that in the lower diagram the distances from your galaxy to all the other galaxies will have increased by this same factor. So whichever galaxy you have chosen, you will see all the other galaxies moving away from yours. As the old quip goes, when estate agents are asked, what are the three most important things for selling a house, the answer is: "location, location, location." In the Universe at large, however, location is not an issue. Whichever galaxy you chose to move into – a desirable giant elliptical galaxy, an elegant spiral, or a dwarf galaxy with limited floor space – you will have pretty much the same view of the Universe, and you will quickly find that most of your neighbors are moving away.

This figure also shows that Hubble's law is exactly what you should expect if the Universe is expanding. Imagine now you are living in galaxy A (a spacious spiral with a bar, with unfortunately some rather irregular neighbors) and look at galaxies B and C. Let us suppose that in the first picture galaxy B is one million light years away from yours and galaxy C is two million light years

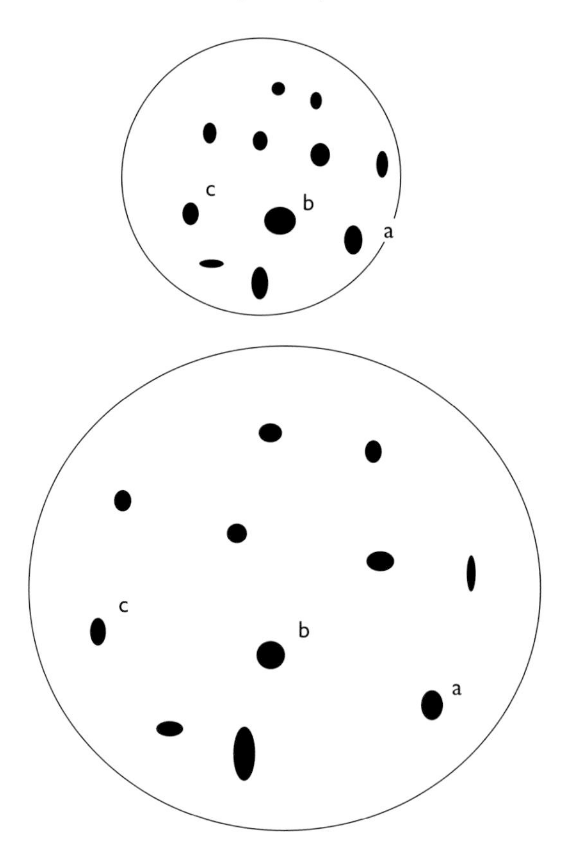

FIGURE 6.5 The upper part of the figure shows a small region of the Universe containing a handful of galaxies (the dark circles and ellipses). The lower part of the figure shows the same region after the Universe has expanded in size by a factor of two.

away. In the second picture, the distance between each pair of galaxies has increased by a factor of two. Galaxy B is now two million light years away and galaxy C is four million light years away. The distance between your galaxy and galaxy B has increased by one million light years, but the distance to galaxy C has increased by two million light years. The average speed of a galaxy is just the distance it has travelled divided by the time taken to travel this distance. In the same interval of time, galaxy C has travelled twice as far as galaxy B. Galaxy C must therefore be moving twice as fast away from you as galaxy B. This is exactly what Hubble found: a galaxy's speed is proportional to its distance from us – a galaxy twice as far away is moving twice as fast.

Hubble's discovery completed the Copernican Revolution. A decade earlier, Harlow Shapley had showed that the Sun is merely an average star in an anonymous suburb of the Galaxy. Now Hubble had shown that even the Galaxy is not special in any way. Our view of the Universe would be fairly similar whichever one of the billions of galaxies we happened to live in.

I find Hubble a mysterious figure. Part of the mystery is that we know surprisingly little about his personality. He lived in a reticent era; he left no memoirs, no personal letters, and had no personal confidant apart from his wife. Despite the paucity of information, his recent biographer, Gale Christianson, has shown that part of his public persona was a lie. He claimed to have practised law and to have fought heroically in the war – Christianson has shown both were untrue. He cut himself off from his Missouri family and also appears to have been something of a bully. Of course, great scientists are not necessarily saints, but in Hubble the gap between the magnitude of the scientific achievements and normal human pettiness seems wider than in most. In later years, Alan Sandage, his one student, had the impression of him that he was "more an actor than a natural patrician"; and with his movie-star looks, his costume – shirt and tie, Norfolk jacket, jodhpurs, high-topped military boots (a mixture of English country gentleman and military officer) – and, most of all, in his elusive personality, he seems to me more like a Hollywood star than a typical scientist.

One thing I do find very appealing about Hubble was his attitude towards theorists. As an observer, Hubble's conservatism was notorious. Long after everyone else was using the term *galaxy*, he was still using the term *nebula*. He remained true to the data, classifying the nebulae and carefully measuring their properties, but sceptical of the theorists' explanations of their significance. By the early 1930s, Hubble was not even convinced that the redshifts he was measuring for the fainter nebulae really did mean the huge speeds implied by the Doppler effect. Most astronomers however were by now convinced, as a result of Hubble's own work and of general relativity, that the Universe is expanding and that space may be *curved*.

We live in a three-dimensional universe, which is the same as saying that we live in a universe in which there are three perpendicular directions: up-and-down, left-and-right, backwards-and-forwards. On the scale of the everyday world space is not curved (or rather any curvature is undetectable), which means that we have no experience to help us imagine the strange effects of curved space. However, a useful way to start thinking about these effects is to lose a dimension and imagine what it would be like to live in a two-dimensional universe.

Imagine that you are Fred, a 2D creature living on the surface of a sphere (Figure 6.6). Your universe is the surface of this sphere, although you are not aware it is a sphere. As a 2D creature, you are aware of backwards-and-forwards and left-and-right, but you are not aware there is an up-and-down. A 3D creature looking at your universe from the outside can see it is curved; but as you crawl around your universe, you are not aware that the sphere has an inside or an outside – these are part of the unimaginable third dimension. Is there any way you could tell you are living in a curved universe?

Apart from travelling right round the Universe, which we will assume is impractical, here is one simple method. Suppose you do some basic high-school geometry: draw a triangle, measure each angle with a protractor and add the three angles together. If you are living in a flat universe, you will find the sum of the angles is

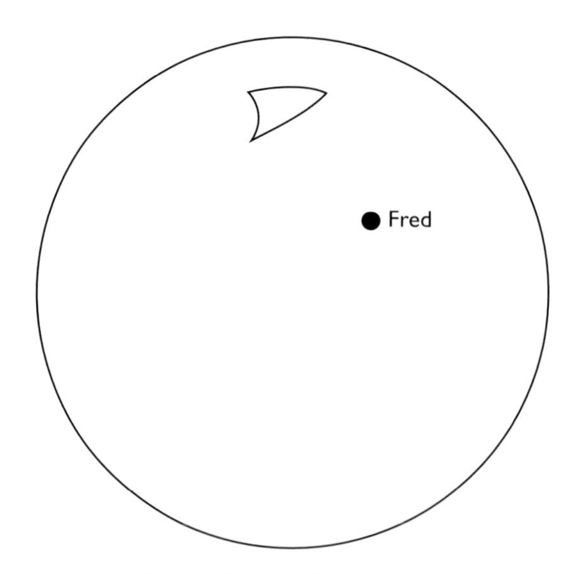

FIGURE 6.6 A universe with only two dimensions.

180 degrees – the standard school result. However, if you are living in curved space, you will find a different value (greater than 180 degrees for the universe in the figure). Whether or not this method would work in practice depends on the size of the triangle. If the triangle is small compared with the size of the Universe, the sum of the angles would be only infinitesimally different from 180 degrees, and it would be impossible to tell anything useful about the curvature. But as long as the triangle is big enough, all that is required for Fred to tell that he is living in a curved universe is some straightforward geometry.

Of course, we could use the same method to investigate whether our own 3D universe is curved. The first person to realize that space might not be flat was the nineteenth-century German mathematician Karl Gauss. Gauss did precisely the same test I have suggested for Fred by measuring the angles of a triangle made by three mountains. He found that the sum of the angles was 180 degrees. This result shows that space *is* flat on the scale of things on Earth, but leaves open the possibility that the curvature of space might be detected on larger scales.

The publication of Einstein's theory of general relativity moved the concept of curved space from the abstract world of mathematics to the real world of physics. In Newton's theory, gravity is a force that acts between two objects across completely empty space, which is rather strange when you think about it. In Einstein's theory, gravity is actually the effect of space itself. The motion of a planet around the Sun is no longer caused by a force between the planet and the Sun, but instead by the curvature of space caused by the presence of the Sun. The planet then moves along the line of least resistance through this curved space, which happens to be an ellipse. Fortunately, there is again a 2D analogy. Imagine a ball rolling across a rubber sheet. If the sheet is flat (a very simple 2D universe), the ball will travel in a straight line, which is what a planet would do in our 3D space if there were no other objects in the Universe. Now suppose that there are objects on the rubber sheet, making hollows in it. If the ball is now rolled across the sheet, it will follow a curved path. *We* can see the distortions in the sheet, but a creature living in this 2D rubber-sheet universe would not be able to see them – he would only know they are there because the ball is not moving in a straight line. In the same way, back in our 3D Universe, the planets' curved paths are caused by the distortion in space produced by the Sun.

The idea of curved space helps to answer an obvious question: does the Universe have an edge? In a universe with an edge, location *would* be important: if we were living in a galaxy close to the edge we would see galaxies on one side of the sky but not on the other. According to general relativity, the Universe is one of a number of possible types, but whichever type it is, it does not have an edge. One possibility is that the Universe, like the 2D universe in Figure 6.6, is curved back on itself. If true, if we set out into space and continued travelling in a straight line, we would end up back at our starting point. The other main possibility is that the Universe is infinite. If the Universe is infinite, space may be flat (the 2D example is a flat sheet of paper) or it may have negative curvature, for which there is no useful 2D example. In either case, if we set out into space and continued travelling in a straight line, we would carry on forever – always surrounded in our travels by the same kinds of galaxies arranged in the same kinds of groups and clusters, but never coming to an edge. The main equation describing these possible universes is called, as a grace note of historical justice, Friedmann's equation.

Of course, Hubble, the ultimate sceptical observer, was not convinced by any of this. In the 1930s, he started a project to test, once and for all, whether the Universe really is expanding, and also whether space is curved. This project is one of the longest running projects in astronomy, because it has continued, in one form or another, for over six decades (Chapters 7 and 8). The basis of the project is to observe objects that are so far away that the effects of curvature, as well as other cosmological effects, become important.

The first part of the project was based on galaxies which Hubble believed might be *standard candles*. To see how this method works, let us suppose we have some real standard candles: candles which are identical in every way. Because all the candles emit exactly the same amount of light, the relative brightness of two candles will only depend on their distances. If one candle is twice as far as another, it will be a factor of four fainter; if one is three times further away, it will be three squared or nine times fainter, and so on. This is the well-known inverse square law between brightness and distance that holds in flat space. However, if space is curved, the relationship between brightness and distance will be different – as long as the objects are far enough away that the effect of the curvature is important. The expansion of the Universe will also alter the everyday relationship, because the time taken for light to travel to us from a distant galaxy is often so long that when the light was emitted the Universe was much smaller than it is today. Hubble realized that if he could find some galaxies that he was confident always emit exactly the same amount of light, he would be able to use these as standard candles to test the conclusion that the Universe is expanding and also to measure the curvature of space. The galaxies he picked were the brightest galaxies in clusters. It was a good choice because these galaxies do all emit roughly the same amount of light, and they have the additional virtue that because they are so luminous they can be seen to a great distance.

The method he used in the other part of the project was the outwardly simple one of counting the number of galaxies on a photographic plate. Hubble had discovered that the longer he exposed his photographic plates, the more galaxies he saw. This is not necessarily so. If the Universe consists of a big clump of galaxies surrounded by empty space, then once the exposure time is long enough to reveal the faintest galaxy in the clump, increasing the exposure time even more will reveal no new galaxies. Hubble however discovered that as he increased the exposure time, the number of galaxies he saw increased in the way one would expect if the Universe is uniformly filled with galaxies (another piece of evidence that location is not important). He realized that if he could increase the sensitivity of his plates even further so that they revealed sufficiently distant galaxies, the curvature of space and the time taken for the light to travel from the galaxies would become important, and the relation between the number of galaxies and the plate sensitivity would begin to deviate from the standard relation.

Hubble spent the rest of his life pursuing this project. Both parts of the project required plates with extremely long exposure times. Since the 100-inch telescope tended to drift away from its position if left, Hubble and Humason, who acted as his assistant, spent night after night in the 1930s sitting in the dark on the telescope platform, looking through an eyepiece and controlling the motion of the telescope with a handheld control pad. Apart from boredom and tiredness, the greatest enemy was the cold. On the coldest nights the observer's tears could freeze his eyelashes to the eyepiece. The faintest galaxies on their most sensitive plates were about ten thousand times fainter than the Andromeda Galaxy. The largest redshift they were able to measure was for a cluster of galaxies in the constellation Ursa Major, a redshift so large that, if it is caused by the Doppler shift, the cluster is moving away from us at 41,600 kilometers each second, about one seventh of the speed of light.

By the mid-1930s, it was clear the project had failed. The plates were not quite sensitive enough and the redshifts not quite high enough for the effects of curvature and light-travel-time to be completely clear. As a result of the construction of the 100-inch telescope, the human horizon had expanded a thousand times, but some of the key effects of the expanding Universe still lay beyond the horizon.

Despite his tremendous scientific achievements, the stream of honors, and the invitations to dinner with movie stars, the last two decades of Hubble's life seem somehow tragic. The way to see beyond the horizon set by the hundred inch telescope was quite obvious: build a larger more sensitive telescope. George Ellery Hale duly obliged, raising the money for a 200-inch telescope to be built on Mount Palomar, well away from the lights of Los Angeles. However, the project was delayed by the Second World War, it took eleven years to grind the mirror into the correct shape, and the telescope was only ready in 1948. Only a year later, Hubble had a heart attack, at the early age of 60, and after that was too ill to make much use of the new telescope. Despite the honours he was given, Hubble seems to have been always anxious for more. During these last years, he was passed over for the directorship of Mount Wilson, probably because of his difficult personality, and he failed to win the greatest scientific prize of all, the Nobel Prize. There is not actually a Nobel Prize for astronomy, but there is one for physics and Hubble was definitely anxious to win it, hiring a publicity agent to work on his behalf. He probably would have won one, but he died, and Nobel Prizes are only given to the living.

There is a photograph of Hubble taken shortly before he died. It must have been taken from the top of the dome of the 200-inch telescope because the camera is looking down through the telescope. The mirror can be seen far below. In the foreground is the prime-focus cage in which the observer sits, suspended in the middle of the telescope by a spider web of metal. Hubble is sitting in the cage, wearing a suit and tie and looking up at the photographer. He is frowning and looks almost frightened of the camera. To me, he looks like a haunted man. However, set against this is a conversation he had with the poet Edith Sitwell one afternoon in his study. He showed her some photographic plates containing galaxies millions of light years away. When she remarked, "How terrifying!" he replied, "Only at first when you are not used to them. Afterwards, they give one comfort. For then you know there is nothing to worry about – nothing at all!"

It seems fitting to give Hubble himself the last word, because if he did not leave a personal memoir, he did write a beautiful book about his exploration of the Universe. At the end of this book he wrote words which will have struck a chord with many observers since his time:

> Thus the exploration of space ends on a note of uncertainty. And necessarily so. We are, by definition, in the center of the observable region. We know our immediate neighborhood rather intimately. With increasing distance, our knowledge fades, and fades rapidly. Eventually, we reach the dim boundary – the utmost limits of our telescopes. There, we measure shadows, and we search among ghostly errors of measurement for landmarks that are scarcely more substantial. The search will continue. Not until the empirical resources are exhausted, need we pass on to the dreamy realms of speculation.

7. The History of Galaxies

It is March 1999. I am on a plane on my way out to use the telescopes at Mauna Kea Observatory in Hawaii. This is a long journey: a very early morning start in Cardiff (so early that people are still leaving the night clubs), the coach down to London; the nine o'clock United Airlines flight to San Francisco – duration 12 hours; the flight from San Francisco to Honolulu – duration five hours; and finally the 40-minute hop from Honolulu to Hilo on the Big Island. It is a journey I do several times a year and all that time on planes away from distractions should be a great opportunity to catch up on work. Somehow I never quite manage it. I pack my bag full of recent scientific papers from my research field which I have yet to read and, in case I just don't feel like science, I put in one of the classics from my bookcase which I have been meaning to read for ages. Usually, however, after half an hour on the plane I am flicking through the channels on the entertainment system and reading the in-flight magazine about what to do in Seattle on a two-day mini-break. The good book remains unread and I read instead the two mysteries I bought in the airport bookstore. When I turn up at the hotel in Hilo twenty seven hours after leaving my house in Cardiff, I am exhausted, and my knowledge of both astronomy and literature is still very patchy.

On this trip I am put to shame by someone. He is a medical doctor, but he did a degree in geology and is still fascinated by the subject. He is sitting next to me, but for him a long trip on a plane is not something to be endured but an opportunity. He takes with him large-scale geological maps and uses them to study the geology of the land over which the plane is passing. He shows me different types of glacier in Greenland and icebergs splitting off the icecap. We talk for the whole journey. He tells me about his passion, geology, and I tell him about my observing trip and how I am carrying out research into the origin of galaxies. As we leave the plane at San Francisco, the old gentleman who was sitting behind us and listening to our conversation tells me that he once heard Hubble give a lecture. While we wrestle our bags down from the overhead bins and line up to leave the plane, I ask him about his impressions of Hubble, but unfortunately he can not remember anything about him.

Seventy percent of all the scientists that have ever lived are still alive. I am not sure whether this statistic is true or whether it is just an urban myth, but it does make some sense. Until the Second World War most scientists were amateurs, and it was only the success of scientists during this war that started the flow of public money that made it relatively easy to make a living as a scientist. Whether or not this statistic is true of scientists in general, it is definitely true that most astronomers are still alive. The reason for this is simple: many types of astronomy and many research fields did not even exist a generation ago. Despite the popular image of an astronomer as being someone who peers up a telescope, nowadays more astronomers use radio, gamma-ray, x-ray, infrared, submillimeter and ultraviolet telescopes than the traditional optical telescope. These are all recent inventions. Radio astronomy is the oldest of these new types of astronomy, yet the first radio astronomer was still alive in the mid-1980s when I heard him give a lecture*. My own research field – trying to understand the history and origin of galaxies – is also relatively new, simply because until Hubble's discovery eighty years ago nobody was sure the nebulae were separate galaxies.

* Strictly, the first person to detect radio waves from outside the Solar System was a radio engineer called Karl Jansky. Grote Reber, whom I heard talk, was also a radio engineer, but he was more interested in the astronomical possibilities of this new waveband. Because professional astronomers were completely blind to these possibilities, Reber in his spare time (he had a full-time job as a radio engineer) built a radio-telescope in his backyard in Wheaton, Illinois, a suburb of Chicago. In the late 1930s he used this telescope to make the first radio map of the Milky Way. He died in 2002.

After Hubble died in 1952, a large part of cosmological* research for the next three decades was essentially a continuation of his program of observing distant galaxies, in order to measure the curvature of space, and more generally to test the conclusion that the Universe is expanding (Chapter 6). Distant galaxies are very faint, and so very large telescopes are needed to collect as much light from them as possible. Until the 1970s, when the new observatories on Mauna Kea and elsewhere were constructed, this program was mostly carried out with the 200-inch telescope on Mount Palomar, Hubble's old telescope. It was also mostly lead by Hubble's former student, Alan Sandage, whom Hubble had taken on when he became aware that he would be too ill to complete his program. Completing Hubble's program became Sandage's life's work, although not without complaint. "If you were an assistant to Dante and Dante died," Sandage apparently once complained to an interviewer, "and you had in your possession the whole of the *Divine Comedy*, what would you do? What would you actually do?"

Ultimately, however, Hubble and Sandage's program failed. The reason for this is, with the twenty-twenty vision of hindsight, rather obvious.

It is a bright spring morning in Cardiff and sunlight is streaming through the window. If I carefully shield my eyes with my hands, I can sneak a look at the Sun. However, the Sun I see does not exist *now*. I am seeing what the Sun looked like eight minutes ago, because it has taken the light that enters my eyes eight minutes to travel from the Sun to me. Of course eight minutes is not very much – the Sun won't have changed very much in this time – but it is enough to show that when we look out into space we are also looking back in time.

The unit often used to measure the distances of stars and galaxies also shows that our perceptions of time and distance are intertwined. Although a *light year* sounds like a unit of time, it is actually a unit of distance – the distance travelled by light in one year. The nearest star, Alpha Centauri, is four light years away, which means that it has taken the light we see four years to reach

---

* Cosmology is the study of the Universe as a whole.

us; and so we are seeing what the star looked like four years ago rather than what it looks like today. It is possible, although not very likely, that the star does not actually exist anymore. If we look further out into the Universe, it is definitely possible that some of the stars we see no longer exist. The center of the Galaxy is thirty thousand light years away, and so we are seeing what the stars there looked like thirty thousand years ago, long before the earliest cities on the Earth.

Outside the Galaxy, the distances and times mount up. The Andromeda Galaxy, the nebula that Hubble showed was a galaxy, is in the Local Group of galaxies, virtually in our backyard; yet it is about two million light years away, which means we are seeing it as it was before Homo Sapiens arose on the Earth. The nearest cluster of galaxies is about fifty million light years away, which means we are looking back in time almost to the era of the dinosaurs. The furthest galaxy whose spectrum Hubble was able to measure with the Mount Wilson 100-inch telescope is much further than this: three billion light years away. His observations therefore did not tell him anything about that galaxy today, but instead what it looked like three billion years ago.

This is the problem that caused Hubble and Sandage's program to fail. The key part of the program was to identify a class of galaxy which always emits the same amount of light. By comparing the relationship between brightness and distance for these *standard candles* seen in the Universe at large with the relationship for standard candles seen in our everyday human-scale world, it should be possible, in principle, to test the expanding Universe hypothesis and also to measure the curvature of space. But what if the galaxies themselves are changing? When we look billions of light years out into space, we are also looking billions of years back in time. What if galaxies in the past were different from those today? By the mid-1970s, when astronomers plotted brightness against distance, they found a bigger difference from the everyday relationship than could be explained by the curvature of space and the expansion of the Universe; they realized that galaxies in the past *were* different – galaxies evolve.

With hindsight, this is not surprising. Galaxies are just large collections of stars, and because stars evolve, galaxies should too. During the Second World War, when the lights of Los Angeles were dimmed due to the wartime blackout, the astronomer Walter Baade used the 100-inch telescope to show that there are two general types of star: stars in *population I* are blue, luminous and short-lived; stars in *population II* are red, dim and long-lived. Baade discovered that elliptical galaxies are mostly made up of population II stars, whereas spirals like our own are a mixture of the two, with the central bulges being mostly population II and the disks being population I. The standard candles used by Hubble and Sandage were ellipticals. Today, ellipticals contain mostly population II stars. Billions of years in the past, at the time when most of the stars in these galaxies were being formed, they would also have contained many short-lived population I stars. Because population I stars are more luminous than those in population II, ellipticals at that time would have been more luminous than they are today.

Although the discovery of galaxy evolution ruined Hubble and Sandage's grand program, it did open up another intriguing possibility: writing a history of galaxies. How have galaxies changed as the Universe has aged? How and why did they form in the first place?

An astronomer has one huge advantage here over an archaeologist trying to write the history of a ruined city. The archaeologist has to infer the history of the city from its debris – the fragments of pottery and the broken tools – and knows that he can never be absolutely sure what really happened. For the astronomer, there is at least the possibility of being absolutely sure, because he does not have to make inferences: he can see into the past. When we look at galaxies at different distances, we are looking at different eras in the Universe's history. By comparing the properties of galaxies at these different distances, it should therefore be possible to watch the change in galaxies as the Universe has aged. It should even be possible, using this cosmic time machine, to see back to the time of creation – the era when the galaxies were formed.

This probably sounds too good to be true. It *is* true, but there is one subtle limitation. Let me try to explain this by the following fantasy.

Imagine that the speed of light is not 300,000 kilometers per second, but is instead only 7 centimeters per hour. This would mean that light would be able to travel a distance of only 613 meters in one year. Let us also imagine that it is possible to look through walls and see around the curve of the Earth. This situation would be a historian's dream, because it would now be possible to look back in time and see what *really* happened in the past. The center of Cardiff is about 4 kilometers from my house, so although with this snail-like speed for light I would be constantly bumping into the furniture, I would also be able to see what was happening in the center of Cardiff about six years ago. This is not particularly interesting, but it gets more interesting as we look further out. London is about 150 kilometers away, so I would be able to see what was happening there about 250 years ago, in the middle of the eighteenth century. If there is one place I would really like to see, it would be Athens in the fifth century BC, a tiny city but packed full of world-famous philosophers, politicians and writers. However, this is where the fantasy breaks down. With the value of the speed of light I have chosen, I would be able to see Italy at about this time, but I would only be able to see what was happening in Greece in about 1200 BC, well before Socrates, Plato and company. The limitation of this form of time travel is that it is possible to see only certain places in each epoch.

There is the same limitation on cosmic archaeology, but it is not as important for cosmic history as it would be for human history. Because astronomers believe that the Universe has always been pretty much the same everywhere (location is not important), it does not really matter that we can only see a small part of it in any era; fifth century Italy and Greece may have been very different, but it does not matter which bit of the Universe ten billion years ago we can see, because that bit should have been the same as any other bit at that time.

The real practical limit on cosmic archaeology is that distant galaxies are very faint. In the 1970s, at the time it was becoming clear that Hubble and Sandage's program was never going to work, cosmic archaeology was not practical. Telescopes were just not sensitive enough.

March 1999: It is the first night of my observing run on the United Kingdom Infrared Telescope at Mauna Kea Observatory. I have just finished dinner and I am being driven by my telescope operator, Thor, up to the summit. We are now about three thou-

sand feet above Hale Pohaku, the astronomers' residence, and the summit, which is not yet in view, is still two thousand feet above us. The ground around us looks like the surface of Mars. There is no vegetation, no soil, only a layer of coarse ground-up brown rock. Everywhere there are signs of past volcanic activity: half-collapsed craters and hills that look odd because there has not yet been time for them to be eroded by the wind and the rain. The last eruption on this mountain actually occurred about a thousand years ago, short on a geological time-scale but long enough ago to make it safe for us. However, if I turn my head, I can see a volcano in action. The surface of the other big mountain on the island, Mauna Loa, which I can see in the distance, is streaked with black lava flows. This lava, which spills into the ocean in a cloud of steam, is stealthily building the island. Fifteen years ago, when I first came here, the island's diameter was tens of meters less than it is today; a few million years ago, the island had not yet emerged from the ocean.

The Ford Bronco powers up the last steep slope, and suddenly we are on the summit of Mauna Kea with its gaggle of a dozen telescopes. The shiny silver dome of my telescope, the UKIRT, is directly in front of us. This is always the moment of truth. Until now the observing run has been a pleasant prospect, a holiday from real life, a journey to a romantic place to carry out the tasks of what has always seemed to me, even in my more cynical moods, a romantic profession. Now there are the very unromantic facts that I have a thirteen-hour working night in front of me, that I am already dead-tired from lack of sleep and jet lag, that I am at 14,000 feet, where the low oxygen level dulls the brain, and that for the next thirteen hours everything will depend on me; Thor will help but all the decisions will have to be mine.

The only way to deal with this is not to think about it. Take one step at a time. Get my bags out of the Bronco. Follow Thor into the control room. Find a space on a table for all my notes and charts; find another space for all the cans of coke, sandwiches, and apples that are meant to get me through the night. Check which object I want to observe first. Wait for Thor to start up the telescope hardware and software and wait for him to give me the word I can start the camera software. Log-on to my computer back in Cardiff to check that there is no vital last-minute e-mail. Start the camera software. Check with Thor about two programs I don't understand. Consult with him about a suitable calibration star. Give him the position of my first target. Check that he has typed the position into the computer correctly, because the altitude makes everyone stupid – telescope operators and astronomers.

Now there is a break. Everything is ready but the Sun has not yet set. I go outside. It is tempting fate to say so, but the weather looks good. There is a thick layer of cloud, but it is several thousand feet below, with the summit an island in a sea of white. It is probably raining in Hilo but here the sky is an uninterrupted blue. Mauna Kea is the sacred mountain of the Hawaiians, but today it too often looks like a building site. Now however the bulldozers are silent and the construction crews have gone home for dinner. In the light of the setting sun, the white and silver of the telescope domes and the blue of the sky look like newly minted colors compared with the tired colors down at sea level.

Immediately to my left, only a hundred feet away, is the white dome of the University of Hawaii's own telescope. One of the smallest telescopes on the mountain, with a mirror only 2.2 meters in diameter, it is also the oldest – and thus has the best site, right on top of the main ridge. Further away, also on the main ridge, is the 3.5-meter Canada–France–Hawaii Telescope, jointly owned by Canada and France, with the University of Hawaii taking its usual gangster's cut of the observing time. The CFHT does not have the biggest mirror on the mountain, but its huge white dome makes it one of the most impressive. Its dome is now open and I can just see the telescope through the slit in the dome. Far beyond and below the CFHT, too far to walk, I can see the squat silver dome of the Infrared Telescope Facility. It is so far away that it is almost a separate observatory, and it is also different from the other telescopes on the mountain because it is funded by NASA and is mainly used to support NASA's planetary exploration program. Behind me and out of sight is Millimeter Valley, the location of the two submillimeter telescopes, the James Clerk Maxwell Telescope, which I have used many times myself, and the Caltech Submillimeter Observatory; both have metal dishes rather than the usual glass mirrors of optical tele-

scopes. Also out of sight from where I am sitting (my bottom is beginning to get cold) are the huge telescopes that have been built in the last few years. Immediately below the main ridge, relegated by their youth to a less choice piece of real estate, there is a line of three of these behemoths: the two Keck Telescopes and the Subaru Telescope.

I wish I was observing on one of these. The Keck Telescopes have mirrors that are bigger than a two-storey house and it's not surprising that most of the big discoveries that have been made in the last few years have been made with them. If I had moved to California when I had the chance, I might be sitting in the control room of the Keck now. UKIRT is a lovely telescope but it is not in the same league as the Keck. When the Gemini Telescope is finished, I guess I'll have access to a really big telescope, but that's too far in the future. I wonder how the construction of Gemini is getting along. It's on the main ridge just behind UKIRT. Can't quite see it from here. I can see the telescope moving within the CFHT dome! It must be dark enough to observe. I must get back to the dome.

In 1996 there was a revolution in the field of galaxy research.

This is my own research field, and as every month a seminal research paper appeared on the Internet, I felt in the uncomfortable position of being a bystander at a revolution. As the year went on, it seemed as if the history of galaxies was being written before my eyes.

The causes of this revolution were three technological advances. The first of these was something I discussed in an earlier chapter: CCD cameras. When the first CCD cameras (CCD, remember, stands for charge-coupled detector) were constructed in the late 1970s, they cost tens of thousands of dollars; today a CCD or digital camera can be bought for less than a hundred bucks. The beauty of CCD cameras for astronomers is their sensitivity. Even the most sensitive photographic plate detects only about one in twenty of the light photons that land on it, but a CCD camera detects virtually every photon. When the first CCD cameras were placed on telescopes, they effectively increased the light-collecting ability of each telescope by over a factor of ten – the equivalent of instantaneously building telescopes with mirrors ten times larger.

The second advance was the construction of even bigger telescopes. Once one has a camera that will detect virtually every photon, the only way to increase the sensitivity of a telescope further is actually to increase the size of its mirror. The desire to build bigger mirrors – to gather more light and so see deeper into the Universe – has of course been one of the wellsprings of astronomical research for over two hundred years, from Herschel to Hale. However, after the 200-inch telescope on Mount Palomar was commissioned in 1948, it was almost five decades before a telescope was built which had a significantly larger mirror*. The reason for this delay was the immense technical challenge of making such a large mirror. Two hundred inches is five meters or about seventeen feet. I am six feet tall, but if I lay across the 200-inch mirror head to toe with two other equally tall men, only one of our heads would be sticking over the edge. The 200-inch mirror was ground from a disk of glass weighing twenty tonnes; the accuracy of the final surface was one millionth of a centimeter; the grinding and polishing of the glass took *eleven* years. Even casting such a large disk of glass was a huge technical challenge, and one of Hale's nervous breakdowns occurred when the initial disk for the smaller 100-inch mirror broke in two.

By the 1980s, astronomers had realized that the way to overcome the problems of making a single large mirror might be to make many small mirrors and then combine them. Using a bequest of 140 million dollars from the businessman William Keck, the University of California and the California Institute of Technology started to build a new large telescope on Mauna Kea. The first of the Keck Telescopes was commissioned in May 1993. It does not have a single large mirror, but when its 36 small mirrors are correctly combined together they effectively perform as a single mirror ten meters in diameter. The area of the Keck mirror, and thus the amount of light it can collect from a faint galaxy, is four times greater than that of the 200-inch. Because of CCDs and

* A telescope with a 6-meter mirror was built in Russia, but the observing site is very poor and the telescope has produced few useful scientific results.

the construction of large telescopes like the Keck Telescope, by the mid-1990s astronomers had at their disposal telescopes that were fifty times more sensitive than those available to astronomers twenty years earlier.

The third of the technological advances was an instrument that now looks as if it belongs in a science museum. Even when it was launched, the Hubble Space Telescope had a somewhat antique appearance. From first design to launch, a space mission can easily take over a decade, which means that by the time the mission flies it often contains out-of-date technology. For the HST, this was exacerbated by two delays. First, there was the Challenger disaster, which led to the suspension of shuttle flights. Then there was the discovery when the telescope was first launched that, unbelievably, its mirror had been ground into the wrong shape. By 1993, when a new instrument (essentially a huge pair of spectacles) was flown up to fix the mirror problem, it had been 14 years and about two billion dollars since the initial design. However, even if the HST now looks like a relic of the Apollo space program, in the 1990s it became a fundamental tool for astronomers wanting to study distant galaxies.

The strength of the HST is not its light-gathering power – it is only a relatively small telescope with a mirror 2 meters in diameter – but its ability to see fine detail. The Earth's atmosphere, which makes stars twinkle and blurs the images we receive from space, limits the amount of detail we can see. For nearby galaxies the Earth's atmosphere is not a serious problem; the Andromeda Galaxy covers an area of sky about the size of the Full Moon and, despite the atmospheric blurring, it is easy to see the galaxy's spiral arms and even pick out, as Hubble did, individual stars within the galaxy. More distant galaxies however cover a much smaller area of sky, and for these the atmospheric blurring is very important. New observatories like Mauna Kea Observatory are built on very high mountains partly because of the desire to reduce this problem by getting above as much of the atmosphere as possible, but even from Mauna Kea the image of a distant galaxy is never much more than an indistinct shapeless blob. The HST, which flies above the atmosphere, is still the only instrument astronomers have available to overcome this problem; the HST will often reveal that the shapeless blob is actually a beautiful spiral galaxy*.

By the 1990s, astronomers therefore had at their disposal a much greater range of tools for studying distant galaxies than they had had in the 1970s. The first big discovery of the revolutionary year, however, was not made with a glamorous new telescope, but with one which had been around for almost two decades, one which was not much bigger than the Mount Wilson 100-inch telescope used by Hubble.

Many astronomers move around the world during the course of an astronomical career, but Simon Lilly has moved around more than most. I have known him for many years, so I can trace his route: Cambridge; Edinburgh; Princeton; Hawaii (where we first met); Toronto, where our paths also crossed; and most recently, since I started this book, back across the Atlantic to Zurich. However, despite this superficial cosmopolitanism, Simon is a characteristically English type, or possibly he is an English type of an earlier generation. Six foot tall, with a small moustache and a public school accent (in Britain, confusingly, public schools are actually private schools), Simon has always seemed to me more like an army officer than a typical astronomer. I can imagine him back in Victorian times, at the high noon of the British Empire, in some remote colony, punctiliously looking after his handful of men and dispensing impartial justice to the natives; then in the evening retiring to his tent to listen to his beloved Wagner. Britain lost its empire and Simon became an astronomer, but some of the talents that would have made him an excellent army officer helped him and his colleagues write the first chapter in the History of Galaxies.

In the early 1990s, Simon was in Canada, working at the University of Toronto. The premier Canadian telescope was then the Canada–France–Hawaii Telescope on Mauna Kea. It has a relatively small mirror, only 3.6 meters across, not much larger than

---

* The HST's ability to see fine detail greatly heightens one's visual appreciation of the Universe. I strongly recommend a stroll through the Hubble Heritage gallery (www.heritage.stsci.edu), which contains the most spectacular pictures taken with the HST.

the Mount Wilson 100-inch telescope, but the advent of CCDs meant that its sensitivity was at least ten times greater than Hubble's telescope. The CFHT also had one supremely powerful instrument. With the 100-inch, Hubble had only been able to obtain one spectrum of a galaxy at a time, but in the 1990s astronomers had begun to build instruments capable of obtaining the spectra of many galaxies simultaneously. The most sophisticated of these *multi-object spectrographs* was on the CFHT. Simon realized that with this instrument and enough observing time on the telescope, it would be possible to carry out an ambitious survey.

The main obstacle was obtaining enough telescope time. As the name suggests, the CFHT is jointly run by Canada and France, with only forty per cent of the time being reserved for Canadian astronomers. Simon could also expect to get only a small part of that forty per cent himself. However, he and another Canadian astronomer, David Crampton, and two French astronomers, Francois Hammer and Olivier Le Fèvre, pooled their observing time, and by doing this they managed to find enough time for the survey. An occupational hazard of scientific collaborations is that they tend to break down in acrimony (scientists are at least as egotistical as anyone else, in my experience). Possibly because of Simon's leadership skills, the Canada–France Redshift Survey appears to have run rather smoothly. By December 1995, when the team published their first results in seven papers in the *Astrophysical Journal*, they had obtained the spectra of 600 faint galaxies.

However, before I describe what they discovered, it might be useful to recall some material from earlier chapters. In Chapter 3, I described the Doppler shift: the fact that the motion of a source of radiation – an ambulance, a star, a faint galaxy – causes a change in the wavelength of the radiation, with the size of the change depending on the speed of the source. I described in Chapter 6 how the Doppler shifts of galaxies are almost always *redshifts*, because virtually all galaxies, apart from a handful of nearby ones, are moving away from ours, and so the spectral lines are shifted to the long-wavelength, red end of the spectrum. The technical definition of the term *redshift* is that it is the change in the wavelength of a spectral line divided by the wavelength the spectral line would have if the galaxy were not moving (a galaxy at rest therefore has a redshift of zero). I also described in Chapter 6 how Hubble discovered that the speed of a galaxy depends on its distance – more distant galaxies are moving away from us faster. The redshift of the galaxy's spectral lines tells us its speed; from the speed, using Hubble's relationship between speed and distance, we can estimate the galaxy's distance; and from the distance and the speed of light, we can calculate the *look-back time* – how far back into the past we are looking. The *redshift* of a galaxy's spectral lines is thus the key to cosmic archaeology.

Simon's team discovered that the spectral lines of many of the galaxies in the survey were shifted by such a large amount that the galaxies had redshifts of almost one. A redshift of one is equivalent to a look-back time of eight billion years. This means that when we observe a galaxy at a redshift of one, we are seeing what it looked like eight billion years in the past (we know *only* what the galaxy looked like eight billion years ago – today the galaxy is too far away for the light to reach us). Back in the 1960s, some objects had been discovered at even higher redshifts, but these objects, the quasars, were not normal galaxies. The light from a quasar is not from stars but from hot gas that is disappearing into a massive black hole. The significance of the work carried out by Simon and his group was that this was the first time anyone had found such a large number of normal galaxies such a long time in the past. The Canada–France Redshift Survey was effectively a census of what the Universe was like eight billion years ago.

This immense gulf of time, much greater than the age of the Earth, made it seem likely that the galaxies found in the survey would look very different from the galaxies in the Universe today. However, when the team observed them with the HST, they discovered galaxies that look quite similar to the ones we see around us (Figure 7.1). There were some differences. A greater proportion of the galaxies were irregulars. There also appeared to be more galaxies with disturbed structures and close companions, suggesting the gravitational force from one galaxy might be distorting the other (I will show a spectacular example of this later in the chapter). But otherwise the first big discovery of the survey was that during the last eight billion years, galaxies did not seem to have changed very much.

They did however discover one very interesting difference. The light from a galaxy is often dominated by the light from a

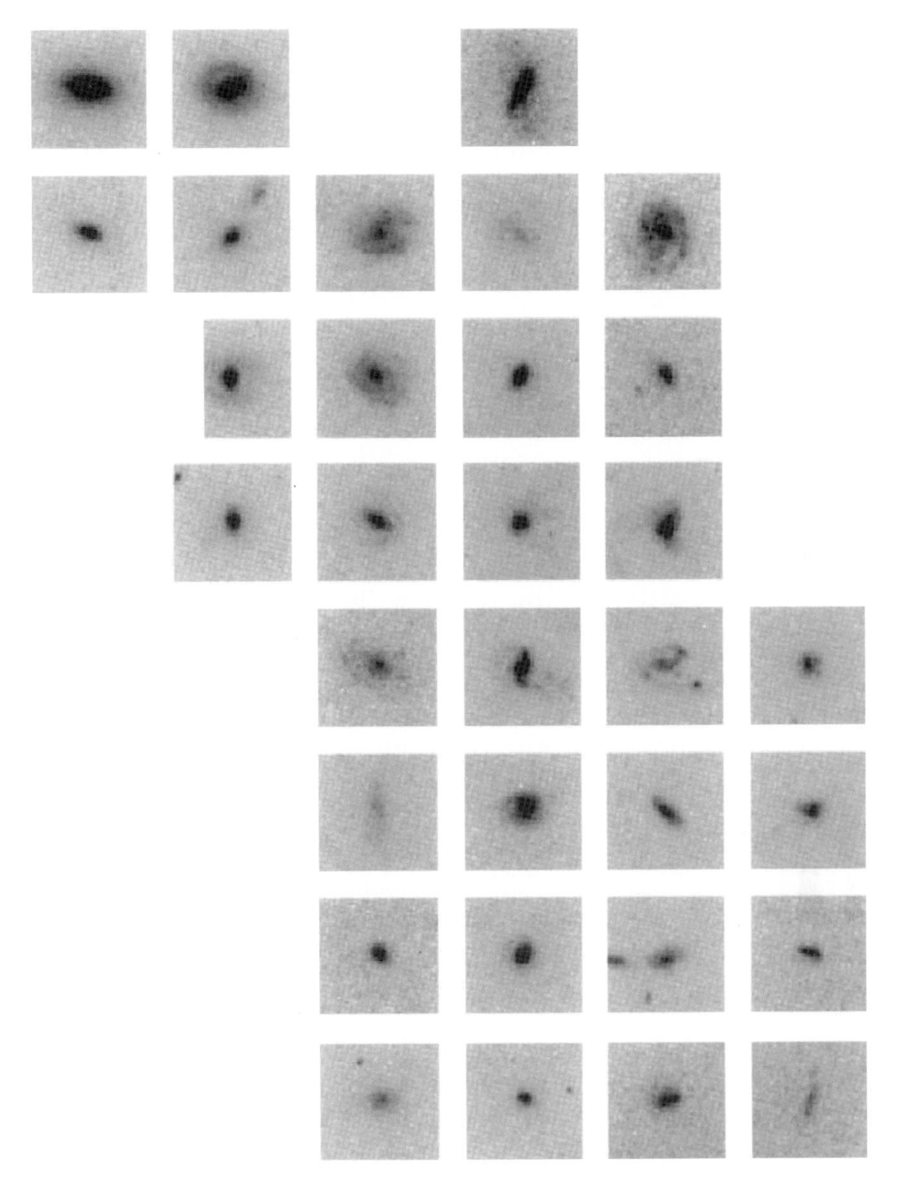

FIGURE 7.1 Images taken with the HST of the galaxies discovered in the Canada–France Redshift Survey. Many of the galaxies look either like normal spirals (the right-hand galaxy in the second row, for example) or ellipticals (see the left-hand galaxy in the first row).

fairly small number of high-mass stars, for the simple reason that high-mass stars have short lives but merry ones; a star with a mass thirty times the mass of the Sun uses its available fuel with such prodigality that it has a luminosity that is almost 30,000 times greater (Chapter 4). The team noticed that although the spiral galaxies in their survey looked quite similar to spirals today, their disks were often much brighter, suggesting that spirals eight billion years ago contained many more high-mass stars. The short spendthrift lives of high-mass stars mean that their number is a good indication of the rate at which stars have recently been forming in a galaxy, and the team concluded that eight billion years ago stars were being formed at a much faster rate in spiral disks than they are today. They extended this calculation by adding up the light from all the galaxies in their survey, which allowed them to estimate the total number of high-mass stars in the Universe at this time. From the number they found, they realized that the star-formation rate in the Universe eight billion years ago must have been over ten times greater than it is today. The vigorous rate at which stars were being formed eight billion years ago shows that, although the galaxies then looked quite similar to those today, the Universe as a whole was a much livelier place.

Only a month after the first results from the Canada–France Redshift Survey were published, the deepest image ever made of the Universe appeared on the Internet.

The HST flies serenely above the atmosphere, orbiting the Earth every ninety minutes, but in one respect it is like all Earth-bound telescopes. Astronomers compete for observing time on the HST by writing a telescope proposal, which is then judged against all the other proposals by a committee of astronomers – a frustrating business because only about 25% of observing proposals are accepted. As with other telescopes, a small percentage of the HST observing time is reserved for the use of the telescope director. Different telescope directors use this *director's discretionary time* in different ways. Some use it for their own research; other, more altruistic directors use it to make replacement observations for astronomers whose observations have been ruined by instrument failure or bad weather; others use it to observe things like supernovae, which happen without warning and are thus difficult to accommodate in the cumbersome time-allocation system. In 1995, the director of the Space Telescope Science Institute, Bob Williams, decided that he was not going to fritter away his discretionary time in the usual mixture of ways, but instead use it for a single important project. Following the advice of an international committee of experts, he decided to use it to obtain the deepest possible image of the Universe.

The astronomers at the Space Telescope Science Institute chose the position for this image very carefully. Since they were primarily interested in distant galaxies, they chose a position well away from the Milky Way with its stars and dust. They also chose a field which was as undistinguished as possible, with no bright galaxies or stars which would make it difficult to see faint galaxies. The observations of the Hubble Deep Field started on December 18th, 1995. For about ten days, over Christmas 1995, the HST stared at a tiny area of sky in the constellation Ursa Major.

Only fifteen days after the observations were finished, the picture of the Hubble Deep Field was shown to the world (Figure 7.2). It immediately became, and has remained, an astronomical icon, the most sensitive picture ever taken of the Universe*.

The HDF covers a minuscule area of sky; if you held a match at arms length, the head of the match would just about cover the HDF. Despite its small size, the image contains over 1000 galaxies and stars. The bright object with the spikes just below and to the left of the center is a star but almost everything else is a galaxy. The faintest galaxies in the image are several thousand times fainter than the faintest galaxies observed by Hubble with the 100-inch telescope and about *four billion times fainter* than the faintest stars in the night sky. Some of the galaxies in the image are normal spirals and ellipticals, but most either have strange irregular structures or are too faint to classify. The black-and-white picture in this book does not show the true beauty of the image*. The best picture of the HDF, the one on every astronomer's wall, is a color picture made from images taken through different colored filters. In this picture, most of the galaxies are blue. Blue is the color of young high-mass stars, which

* This is actually not true, because there is now an even deeper image – the Hubble Ultra-Deep Field – but it is unlikely that this image will ever replace the original in the hearts of astronomers.

* A link to the color picture can be found on this book's website (www.originquestions.com).

FIGURE 7.2 The Hubble Deep Field. Credit: R. Williams (STScI), the Hubble Deep Field Team and NASA

suggests again that early in the history of the Universe stars were being formed much more rapidly than they are today.

As a text for revealing the history of galaxies, however, the picture of the HDF was initially about as useful as an ancient manuscript written in an unknown language. The reason was that at the time of its release almost no redshifts had been measured for any of the galaxies in the image. Without a redshift, a faint galaxy might be a nearby dwarf galaxy, which is faint because it contains relatively few stars, or it might be a huge galaxy billions of light years away, which is faint because it is so far away. The scientific truths in this beautiful picture have only gradually emerged over the last decade as astronomers have measured redshifts and observed the HDF with even more exotic telescopes than the HST (such as the Chandra X-ray telescope and Spitzer Space Telescope, which are both orbiting the Earth as I write).

Only a month after the release of the HDF image, another draft chapter for the History of Galaxies appeared on the Internet. From 1920 onwards, for about five decades, California, with its sunshine, its clear night skies and its huge telescopes built with the money from private benefactors, was the place to be for an astronomer. Then for two decades the action moved elsewhere, as government research organizations built telescopes that were no bigger than those in California but were on better sites and with better instruments. But in the 1990s, Californian astronomers once again had access to the best optical telescope in the world: the 10-meter Keck Telescope.

For an ambitious young astronomer in the early 1990s, however, the place to go was not just California, but a tiny institution in Los Angeles: the California Institute of Technology. Although some other American universities – Harvard, Yale, Princeton – may be better known to the public, for a scientist Caltech is the center of the Universe. As one strolls across the lush lawns of its small campus, dotted not with the brutal concrete blocks of many university campuses but with small elegant Spanish-style buildings, it is impossible not to think of the famous scientists that have strolled over the same lawns. Hubble lectured here; Einstein spent time here; Richard Feynmann spent most of his career at Caltech; a Caltech astronomer discovered quasars; antimatter was first created here in the 1930s; the nature of the chemical bonds holding all matter together was first elucidated at Caltech; quarks were invented on a Caltech blackboard – the list of Caltech firsts goes on and on. Access to the Keck Telescope is shared by Caltech and the University of California, but the university is a huge institution of many sprawling campuses and its observing time has to be shared between hundreds of scientists. In the early 1990s, if one wanted a big slice of observing time on the Keck, the place to go was Caltech. Almost as soon as the Keck Telescope was opened, a young astronomer at Caltech, Chuck Steidel, used it to extend the History of Galaxies even further back in time.

The key to his method was that within the splendid diversity of galaxies, they all have one thing in common: they are all virtually invisible at a wavelength below 912 nanometers (one meter contains one billion nanometers, remember). The reason for this is that hydrogen, the most common element in the Universe, absorbs virtually all radiation with wavelengths less than this. This is an interesting fact, but it is not usually an important one for observers, because this critical wavelength, which is called the *Lyman break*, falls in the ultraviolet part of the spectrum and the Earth's atmosphere is anyway opaque to ultraviolet radiation.

However, suppose there is a faint galaxy in a deep CCD image which actually has a redshift of three. According to the definition of redshift that I gave above, the spectral lines would be shifted in wavelength by a factor of four. The look-back time of a galaxy at this redshift would be about 12 billion years. The Lyman break of the galaxy would now no longer be in the ultraviolet waveband as we observe things on Earth. The radiation which was emitted by the galaxy at 912 nanometers, in the ultraviolet waveband, would be detected by the telescope 12 billion years later at 3648 nanometers – in the visible waveband.

Now suppose that actually we have obtained two deep images, one through a filter which lets through light with wavelengths greater than 3648 nanometers and one through a filter which lets through light with wavelengths less than this. The galaxy would be visible in the first image but would vanish in the second, because this filter only lets through wavelengths below the Lyman break. Suppose that among the many other faint galaxies in the images there are some others that vanish in the second image – these too might be at a redshift of about three. I do not know who first suggested this, but by the early 1990s this idea for finding high-redshift galaxies had been circulating within the astronomical community for several years.

The reason this method had not been used in practice – the reason why Chuck Steidel had a big advantage over other young astronomers in the early 1990s – is that the only way to be absolutely sure about the redshift of a galaxy is to observe its spectral lines. At that time, the only telescope with a big enough mirror, and thus the sensitivity necessary to detect the spectral lines from such distant galaxies, was the Keck Telescope.

Steidel and his collaborators began to use the Keck Telescope to measure the redshifts of the 'drop-out' galaxies, ones that were visible in one image but disappeared in a second image taken at shorter wavelengths. By February 1996, when their first paper appeared on the Internet, they had a sample of 20 galaxies which their Keck observations had shown were definitely at a redshift of close to three. As I write, over 1000 of these *Lyman-break galaxies* have been discovered. We are now looking back in time 12 billion years. I want to emphasize that we are not *inferring* what the Universe was like at this time: we are *looking* in exactly the same way I am looking at this computer screen, the only difference being that the light has taken rather longer to get to us. The latest results from the WMAP satellite (Chapter 8) imply that the Universe began 13.7 billion years ago. And so when we observe a Lyman-break galaxy we are looking back to a time only about two billion years after the Big Bang.

For cosmic archaeologists, the Lyman-break galaxies were another crucial find. The Canada–France Redshift Survey revealed that galaxies eight billion years ago looked similar to those today. But what did galaxies look like four billion years earlier than this?

When Steidel's team obtained HST images of the Lyman-break galaxies (Figure 7.3), they found that things did appear to have changed in the intervening four billion years. Twelve billion years ago, galaxies did not look anything like the spirals and ellipticals we see around us in the Universe today, and which seem to have been present in the Universe even eight billion years ago. The galaxies twelve billion years ago seem to have been quite small, resembling fragments of galaxies rather than the big galaxies we see around us.

The HST images also revealed that sometimes what appears to be a single Lyman break galaxy when observed from the ground is actually a pair of galaxies. Often one or other of the galaxies appears to have a peculiar structure, as if it is being distorted by the gravitational force of the other galaxy. This was an important discovery because theorists have for a long time believed that gravity is the main cause of galaxy evolution. According to the theorists, the first objects that were formed in the Universe shortly after the Big Bang were quite small objects, about the size of globular clusters, and it was only gravity and fourteen billion years

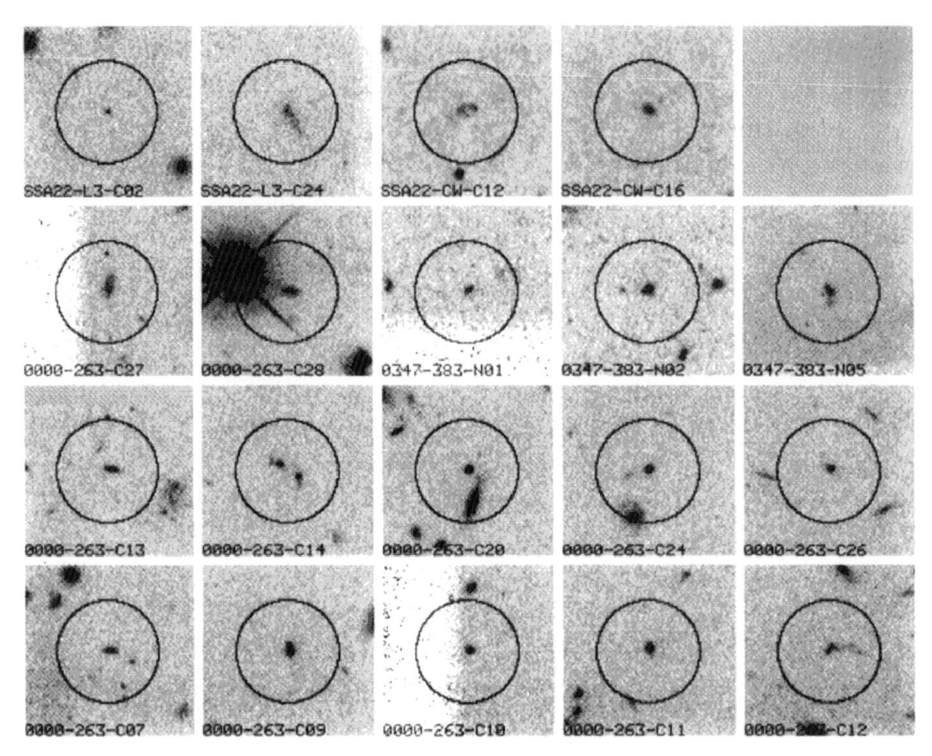

FIGURE 7.3 HST images of the Lyman-break galaxies. There are several images where it looks as if two galaxies may be interacting (see, for example, the third image from the left in the top row and the second image from the left in the third row)

that caused the small objects that formed first to gradually merge together into the big galaxies we see around us today. The History of Galaxies that was beginning to appear in the first months of 1996, although it still had many huge gaps, agreed quite well with the theorists' preconceptions. The observers were finding small objects early in the history of the Universe and only big galaxies later on. They were also finding signs in their HST images of the importance of gravitational interactions between galaxies.

Some of the images in Figure 7.3 do look as if a galaxy is being distorted by the gravitational force of a nearby galaxy, but it must be admitted the images are not completely convincing – even the HST does not reveal much more than a couple of smudges. Figure 7.4 however shows a much more dramatic example of a pair of interacting galaxies: the nearby galaxies NGC 4038 and NGC

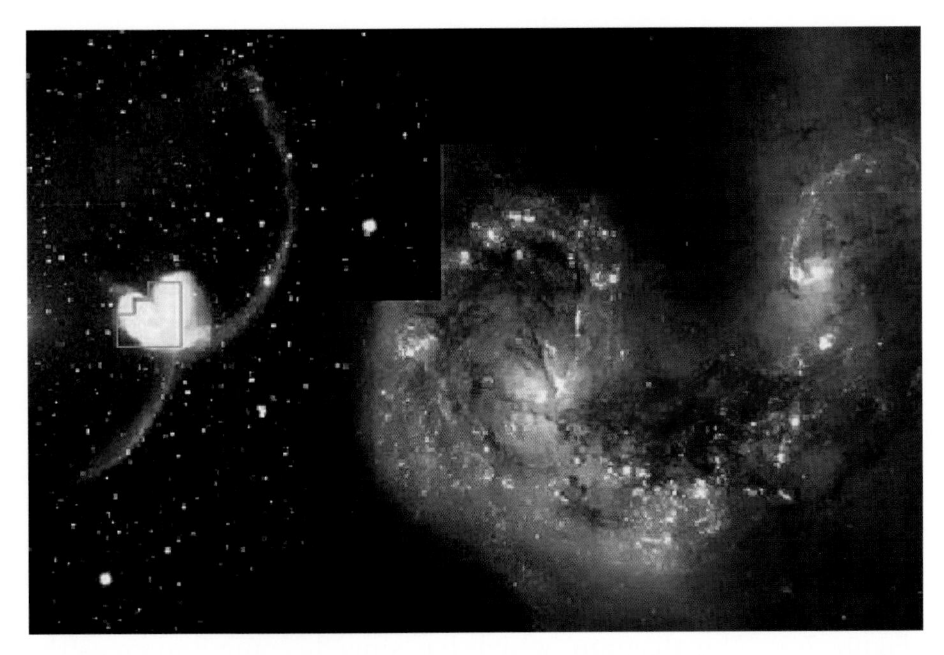

FIGURE 7.4 Two pictures of the Antennae. The left-hand picture was taken with a telescope on the ground. The right-hand picture was taken with the HST and is of the region outlined in the left-hand picture. Credit: Brad Whitmore and NASA/STScI

4039, called for an obvious reason the *Antennae*. The antennae are rivers of stars which have been drawn from each of the galaxies by the gravitational force of the other galaxy. These rivers are only one of the beautiful effects which can be sculpted by gravity. In some interactions gravity builds a bridge of stars between the galaxies; in others, after one galaxy has passed completely through the other (galaxies are mostly empty space), the latter is left surrounded by a ring of stars. There are two possible outcomes of the gravitational dance of two galaxies. Sometimes one galaxy will flirt with another, swinging past it, drawing out rivers and bridges of stars, but ultimately retreating into the darkness of space. Other times the two partners will become locked together. The two will dance together in ever-decreasing circles, gravity disrupting each galaxy more and more, until they merge into a single object. This incidentally may be the future of our Galaxy. We are currently heading towards the other big galaxy in the Local Group, Andromeda, and it seems likely that the two will eventually merge. Some theorists claim that when two spirals merge, the galaxy that is formed is an elliptical, so our descendants in the far future may well find themselves living in a different kind of galaxy.

The Italian astronomer Piero Madau used the results of Steidel's team to make another important discovery. As Lilly's team had done earlier, he assumed the light from a Lyman-break galaxy is dominated by the light from the high-mass stars. He estimated the number of high-mass stars in each galaxy from its brightness, which immediately told him, because high-mass stars have very short lives, the recent star-formation rate in that galaxy. He then estimated the star-formation rate in the Universe 12 billion years ago – as Lilly's team had done for the Universe eight billion years ago – by simply adding up the total number of high-mass stars in all the Lyman-break galaxies. Putting his estimate together with the results of Lilly's team, he plotted a diagram showing how the star-formation rate in the Universe has changed during its history (Figure 7.5). This diagram, which has since become known as the

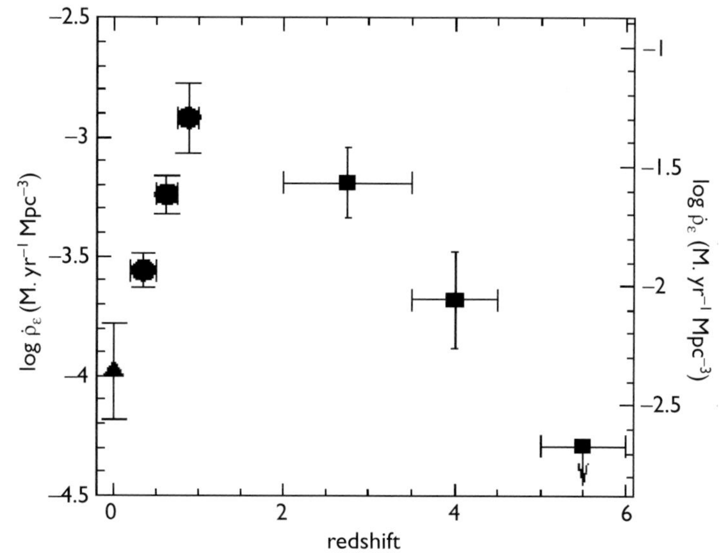

FIGURE 7.5 The history of star formation in the Universe. Redshift is plotted along the horizontal axis. The quantity plotted along the vertical axis is a specialized astronomical term, but effectively it is the star-formation rate in the Universe at that redshift. There is an increase in the star-formation rate by a factor of ten from a redshift of zero (the Universe today) to a redshift of one (the Universe eight billion years ago). Credit: Piero Madau

Madau diagram, shows that although the star-formation rate eight billion years ago was ten times greater than it is today, the star-formation rate at still earlier times was actually less than it was eight billion years ago. Theorists have always liked the Madau diagram, because their simulations suggest that about eight billion years ago is when galaxy-building, the merging of small galaxies to make bigger galaxies, would have been at its peak. These mergers would almost certainly have led to the formation of many new stars, because when two galaxies collide the gas clouds in them will also collide (the points of light in the HST picture of the Antennae in Figure 7.4 are almost certainly new star clusters formed as a result of the collision). Many observers however were surprised by the Madau diagram, for a reason I will describe in a moment.

March 1999: I am relaxing in the UKIRT control room. The rush of setting up the instrument and carrying out the calibration observations is over. I have observed a bright calibration star and the telescope and camera seem to be working fine. Now I am making my first real observation. Forty five minutes to wait until the exposure is finished. Nothing to do! I could check my e-mail, but there does not seem much point as it is only six in the morning in the UK. Thor has just walked in. With his long silver hair, he does look a bit like the god of thunder. What a strange job spending a third of your life on top of a mountain! It's not surprising that telescope operators have more interesting personalities than astronomers. (Do you hate the monotony of a nine-to-five job? Would you like to spend your life in an exotic resort close to a ski slope with satellite television and an interesting international cuisine? Can you combine the roles of engineer, software consultant, chauffeur, weather forecaster and therapist? Then we have the job for you.)

The forty five minutes are up! Quick, give Thor the position of the next target and start another observation. Let me have a look at what I have got. The infrared image on the screen looks pretty good – many faint galaxies. But is there anything present at the position of the submillimeter source I have discovered with the new camera on the JCMT – SCUBA? I wish the position of the submillimeter source was more accurate. There is something present in the image – two faint smudges which look like a pair of distant colliding galaxies – but given the uncertainty in the position it's hard to be sure they are connected to the SCUBA source. There is nothing I can do now. I'll just have to think very carefully about this when I get home. There are thirty five minutes until the next observation is finished. Perhaps I should check my e-mail.

After midnight time blurs or perhaps I just keep falling asleep. By six o'clock I have infrared images of ten submillimeter sources. The sky is beginning to brighten. I make a last observation of a calibration star and then we close down the telescope. As we leave the dome, the weather is still fine; the Sun has just risen and the summit is an island in an ocean of cloud stretching to the horizon – a successful night.

Half an hour later, I am sitting over breakfast in Hale Pohaku. Observers from other telescopes drift in. Nobody talks very much; we are all too tired, except for Thor and the other telescope operators, who are sitting at a separate table chatting and reading the newspapers – to them this is all in a day's work. I look out through the window at Mauna Loa, a lava-streaked swollen lump filling the view from horizon to horizon, bigger from base to summit than Everest, although half of it is hidden under the Pacific. The weather on the two summits is almost always the same. Below me, in the valley between the two mountains, there is cloud, but above Mauna Loa there are no clouds at all, just a few wisps around the summit. The weather should be good tonight.

Elliptical galaxies were the reason why many observers were surprised by the Madau diagram. As Walter Baade had discovered during the wartime blackout, ellipticals are composed almost entirely of Population II stars – red, dim, and very old. It is the age which creates the problem. If the stars in a nearby elliptical galaxy are twelve billion years old, as some estimates suggest, twelve billion years ago stars must have been forming in that galaxy very quickly. This implies that ellipticals twelve billion years ago must have been very luminous because they would then have contained many short-lived luminous high-mass stars (twelve billion years later, only the dim long-lived ones are left). For several decades, astronomers had expected that, if they only looked far enough out into the Universe (and thus far enough back in time), they would be able to see these young elliptical galaxies – beacons of light on the cosmic horizon. By 1996, astronomers were already quite surprised that these objects had not been discovered. The Madau diagram just made things worse, because it implied that the star-formation rate in the Universe twelve billion years ago was not actually very high.

There was one obvious suspect in this mystery: interstellar dust. Back in the 1980s, several astronomers had suggested that very young elliptical galaxies, *proto-ellipticals*, might contain large amounts of dust, which would explain the failure of optical surveys to find them. There was also the possibility that the Madau diagram might be seriously flawed because of dust. If galaxies twelve billion years ago contained more dust than galaxies today, much of their light would be hidden by the dust, and stars might actually be forming in them more quickly than Madau had estimated.

The next chapter in the History of Galaxies was written in April 1996. In 1989 NASA had launched a small astronomy satellite, the Cosmic Background Explorer, or COBE for short. COBE cost only about one tenth as much as the HST, but this small satellite has arguably had an even bigger impact on astronomy (I will describe COBE's biggest discovery in the next chapter). As I explained in Chapter 5, one way to tell that dust is absorbing visible light is to detect submillimeter radiation from the dust, which shows that something is heating the dust. In the late 1980s it was still difficult to detect submillimeter radiation from even nearby galaxies. Rather than trying to detect submillimeter radiation from individual galaxies, one of COBE's goals was to measure the combined submillimeter radiation from all the galaxies in the sky – the *background radiation*.

This was still however a very demanding goal, because the Solar System and the Galaxy contain plenty of dust, and the submillimeter radiation from this dust is much stronger than the submillimeter radiation from the galaxies beyond. By 1996, the NASA team had spent seven years struggling with the problem of removing the effect of this nearby dust from the data. The team took a bit too long and in 1996 they were scooped. A French group, led by Jean-Loup Puget, had been independently analyzing the data from the satellite, and in April this group announced that they had detected the submillimeter background radiation. The result was exciting because the radiation was unexpectedly strong. The French group discovered that the amount of energy emitted by galaxies in the submillimeter waveband is as great as the amount of energy they emit at visible wavelengths.

The importance of this discovery was that it was the first proper accounting of the effect of dust in the Universe. We have always known – not least from the many beautiful pictures caused by dust (Chapter 5) – that dust is important in individual objects. The French team's discovery showed that dust has a huge effect on the Universe as a whole. Their measurements showed that approximately half the visible light emitted since the Big Bang by all the objects in the Universe – stars, galaxies and quasars – has been absorbed by dust; with the stolen energy then being laundered and re-radiated in the submillimeter waveband. Their discovery demonstrated that for anyone wanting to write a history of galaxies, it is not enough to take pretty pictures with the HST, because any optical image only contains half the energy. Any serious cosmic archaeologist has to worry about dust.

This was the last event of a remarkable year. In this short period of five months, astronomers had gone from only being able to speculate about the history of galaxies to being able to observe this history directly – to literally see history as it happened.

The next story I want to tell is the only one in this book in which I was directly involved. In Chapter 5 I described the opening up of the final electromagnetic frontier – the submillimeter waveband. A key moment in the domestication of this frontier was the installation of the world's first submillimeter camera, SCUBA, on the James Clerk Maxwell Telescope in 1997. In that chapter I described the contribution that SCUBA made to the study of protostars, but it was even more important for astronomers in my own research field. Protostars are quite strong submillimeter sources and, as I described in Chapter 5, the first Class-0 protostar was actually discovered before this. However, before SCUBA virtually nothing was known about the submillimeter sources outside the Galaxy, and so its installation offered the prospect of making some interesting discoveries.

Of course, everyone knew this. An astronomer had only to remember quasars, which were discovered in the first radio surveys, to be anxious to use SCUBA. As the "first light" of

SCUBA approached, astronomers frantically formed teams they thought might impress the committees that allocated the time on the JCMT.

The makeup of my own team was a matter of chance and geography. From 1990 to 1994, I worked at the University of Toronto, where I hooked up again with Simon Lilly, whom I had first met when we were both working at the IFA in Hawaii (between Hawaii and Toronto I had spent a year at the Space Telescope Science Institute in Baltimore). This was well before SCUBA was installed on the JCMT, but it was already under construction, and we both realized its potential importance. We agreed that when SCUBA was finally ready, we would apply for observing time to carry out a large survey. The basic goal of our survey would be the obvious one of discovering what submillimeter sources are out there in the Universe. But in the back of both our minds was the realization that if proto-ellipticals are hidden by dust from optical telescopes, they must be emitting submillimeter radiation.

In 1994 I moved to a permanent astronomy job in Cardiff. Three years later, when it became possible to apply for observing time with SCUBA, I applied for time through the British time allocation committee and Simon applied through the Canadian committee. We were both successful, and by pooling our observing time we had enough for the survey, which we decided to call the Canada–UK Deep Submillimeter Survey. Our first allocation of time was seventeen nights in November 1997. This was the first big allocation of time with SCUBA, and I remember feeling quite smug because it put our team in the pole position in the race to make the big discovery. Simon flew out to Hawaii from Toronto and I and my student Loretta Dunne flew out from London.

The instant we arrived on the summit of Mauna Kea, SCUBA broke. The SCUBA instrument scientist, Wayne Holland, told us that it would not be ready to use again for several weeks. To make matters worse, the team due at the telescope after it was repaired would be our main rivals. And to make matters even worse, I realized I had bought a cheap air ticket which could not be changed, which meant I was faced with the prospect of spending two weeks on Mauna Kea, miserable and with nothing to do.

Thanks to a nice woman on the United Airlines counter in Honololu, I did manage to get home, but in January our main rivals went out to use the telescope. I knew they had had excellent weather and that SCUBA had worked well for them, but I just about managed to keep up my spirits because we had some more observing time two weeks later.

Simon had some other commitments and could not go out to Hawaii. I got as far as Heathrow Airport. When I stepped out of the coach at the airport, I heard my name on the public address system. There was a message to call home. It was a gray January day, the rain was sluicing down, and Heathrow airport is not a very attractive place at the best of times. I called my wife on a payphone at the side of the road, while being whipped by the spray from passing cars. She told me Wayne had called from Hawaii and said that SCUBA was broken again. I went back home, this time resigned to the fact that we had lost the race.

The next morning when I checked my e-mail there was a message from Wayne. He had made a mistake. SCUBA was not broken after all. Unfortunately, there was now no way I could get out to Hawaii for the observing run.

Good luck finally intervened. Loretta had already reached Hawaii and she managed to complete our initial set of observations. The weather was superb, as it was for all the rival teams during the first few months of 1998. By a fortunate coincidence, just at the moment when excellent conditions were really needed because of the possibility of using this revolutionary camera, the air above Mauna Kea was exceptionally dry. In the ten years since the construction of the JCMT, the weather had never been as good; it has never been as good since.

At this time, none of the rival teams had any idea whether they had discovered anything, because there was a complex sequence of analysis programs which had to be applied to the data before we saw a submillimeter picture of the sky. During January and February in 1998 the rival teams frantically analyzed their data. For any observing program the moment of truth is when the first image appears on the computer screen. For the SCUBA surveys, this moment would be more exciting than usual because none of us had any idea what to expect.

The moment of truth for me occurred in late February 1998 in my office back in Cardiff. I was alone. I had just applied the last of the SCUBA analysis programs and the first submillimeter image appeared on my computer screen. There was a blob in the top right-hand corner of the image. There were some high-redshift galaxies already known in this part of the sky because we had carried out our SCUBA survey in one of the same fields that Simon and his team had observed earlier as part of the Canada–France Redshift Survey. I had no idea what a real submillimeter source would look like, but when I looked in Simon's catalogue of high-redshift galaxies, I found one at almost exactly the position of the blob. At that moment, I knew our survey had been a success.

With the hindsight of half-a-dozen years, all the competition seems faintly ridiculous. Many different teams carried out surveys of different parts of the sky using a variety of different techniques. All made the same discovery at roughly the same time. The image that illustrates this discovery best was produced by the team observing before us in January 1998. This team, which was led by the British astronomers Jim Dunlop and Michael Rowan-Robinson, used SCUBA to make a submillimeter image of the Hubble Deep Field. Figure 7.6 shows their SCUBA image alongside the HST image. The HST image is incomparably prettier; there are over 1000 objects in the HST image compared with only five blobs in the SCUBA image. But the SCUBA image reveals something new about the Universe. At the position of the brightest of the SCUBA blobs, there is nothing present at all in the HST image. This is a galaxy that is so hidden by dust that it can not be seen at visible wavelengths, even in the most sensitive optical image ever made. It is also emitting so much submillimeter radiation that it is roughly 200 times more luminous, when the radiation emitted in all wavebands is added together, than the galaxies visible in the optical picture. The SCUBA surveys revealed that billions of years ago there was a population of luminous dust-enshrouded galaxies whose existence was not even suspected until the advent of submillimeter astronomy. Among the observers, the gold medal for the discovery should probably go to three young British astronomers, Andrew Blain, Rob Ivison and Ian Smail, who used SCUBA the previous summer, the first month it was on the telescope, to observe some nearby clusters of galaxies. They found the same objects behind the clusters that the other groups found

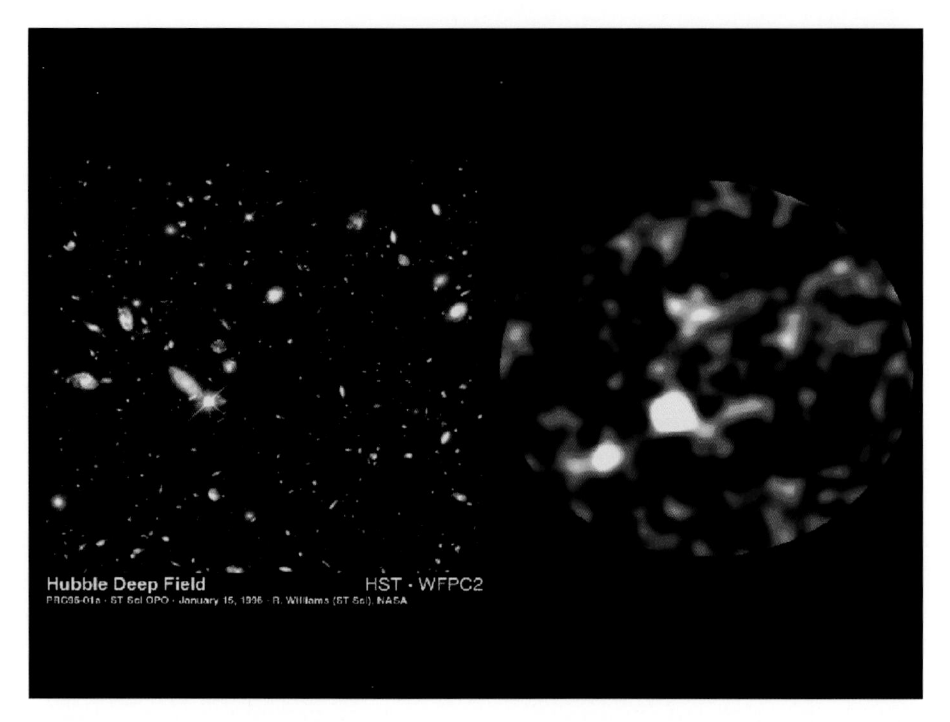

FIGURE 7.6 Two alternative views of the Universe. On the left is the Hubble Deep Field, the deep optical image made with the HST. On the right is the submillimeter image of the same area of sky made with SCUBA. Credit: Jim Dunlop

later that winter. The real winners however were not the observers but the scientists and engineers who built such a revolutionary camera.

Over the last half-dozen years, I and many other astronomers have expended a lot of telescope time in trying to understand these objects. One of the big problems is that these objects are extremely faint – in every waveband except the submillimeter – and so require very long observations. The dust makes them particularly faint, of course, at optical wavelengths, and so measuring their redshifts has been a major challenge. Last year, a team of astronomers led by Caltech astronomer Scott Chapman, by making long exposures with the Keck Telescope, finally succeeded in measuring the redshifts of a significant number of SCUBA sources. These typically lie between 2 and 3, and so when we observe these objects we are looking a long way back in time – between 10 and 12 billion years.

An obvious question is: what is the cause of the phenomenal luminosity of these objects? A little bit of visible light trickles out, but ninety nine per cent of it is absorbed by the dust. One possibility is that the dust is hiding a quasar, an extremely luminous object that contains a massive black hole. However, this now seems unlikely because quasars are strong x-ray sources, and neither the Chandra nor XMM–Newton x-ray observatories have detected strong x-ray emission at the positions of the SCUBA sources.

It now seems almost certain that the thing that is heating the dust is starlight. Astronomers have estimated the star-formation rates in these galaxies from the total amount of radiation they emit, in the same way that optical astronomers estimated the star-formation rates in Lyman-break galaxies. The rate at which stars are currently being formed in our Galaxy is not very impressive: about three stars of the mass of the Sun every year. In the SCUBA galaxies, however, the amount of gas turned into stars every year is about 1000 solar masses, over 300 times greater than in our own. A star-formation rate of 1000 solar masses per year may still not seem very impressive, but this is enough to make all the stars in a galaxy like our own in 100 million years – a cosmological blink of the eye in today's 14-billion-year-old Universe.

Dust-enshrouded objects billions of years in the past in which stars are being formed at a prodigious rate – these are exactly the properties expected for the ancestors of elliptical galaxies. Most observers studying these objects now believe the SCUBA galaxies *are* proto-ellipticals. It seems difficult to avoid the conclusion that an elliptical starts its life as an object full of gas, dust and young stars; that the gas and dust are gradually consumed by the birth of more stars; and that eventually, after ten billion years or so, when all the gas and dust are gone, an object like one of today's elliptical galaxies remains.

No theorist predicted the existence of the SCUBA galaxies. The most widely accepted theoretical model, the bottom-up model, in which small galaxies gradually merge to form big galaxies, predicts that galaxies with large masses should only be found fairly late in the history of the Universe. The big problem for the theorists is that the SCUBA galaxies are being seen early in the history of the Universe but their high luminosity suggests they are massive galaxies. Things are getting uncomfortable for the theorists, but at the moment they have a get-out-of-jail card. The SCUBA galaxies are very luminous, but it is not yet certain, because there is no easy way to measure this, that they also have high masses.

We have now reached Hubble's "dim boundary – the utmost limits of our telescopes." The last decade has produced a draft of the *History of Galaxies*, but there are few details and many gaps, and it will be impossible to turn this draft into an authoritative history with current telescopes. As in Hubble's day, when he exhausted the possibilities of the 100-inch telescope, the answer lies on the horizon in the form of new telescopes. As I described in Chapter 5, the ultimate submillimeter telescope will be the Atacama Large Millimeter Array, an array of 64 dishes which will be located in the Atacama Desert, high in the Chilean Andes. With ALMA, it will be possible to measure the speed of the gas in a SCUBA galaxy as it orbits around the center of the galaxy, and so measure the galaxy's mass. The James Webb Space Telescope is the step beyond the HST. The JWST will have a mirror six meters in diameter. This is too big to fit in the Space Shuttle, so the telescope will be flown into space with the mirror folded up; when the telescope is unloaded the mirror will open up like the petals of a flower. Whenever there has been a problem with the HST it has been possible to send up astronauts to fix it. This will not be possible with the JWST. To maximize the sensitivity of the telescope, it will be sent deep into space – if anything goes wrong out there, that will be that. The main goal of the JWST will be to observe galaxies during the first two billion years after the Big Bang, a period which is currently missing from the history.

This chapter has been about looking back in time. As astronomers fly out to telescopes, make their observations, fly back home, analyze their results, write papers, which drop like stones into the scientific ocean, some making ripples others not, real time drifts onwards. Scientific research seems to come in waves. During the second half of the 1990s astronomers working on the History of Galaxies were passing through the crest of a wave. Now we are in a trough. ALMA and the JWST are in the future. The next wave is on the horizon.

Part IV
The Universe

8. Watching the Big Bang on Television

Anyone who wants to make a television program about the Big Bang faces a big problem. How do you show the Big Bang on TV? The solution TV producers usually stumble across is the same as the one adopted by the designer of the Exhibition of the Evolving Earth (Chapter 1).

In the beginning, the screen is completely dark. A point of light suddenly appears in its center; light begins to stream away from the center; swirling clouds of gas become visible; the clouds of gas begin to form into galaxies . . .

The trouble with this TV version of the Big Bang is that it, and even the words *Big Bang** themselves, creates a deceptive picture of the beginning of the Universe. The main thing that is wrong is that both the words and images give the impression that the Big Bang was simply the explosion of a big lump of matter in the middle of otherwise empty space. The first thing that is wrong with this impression is that the Universe began at the instant of the Big Bang, so there could not have been a lump of matter or even empty space before that instant. Indeed, the word *before* only has any meaning if there is a sequence of events in time, so as time began at the instant of the Big Bang, there was nothing, no space, not even a thought or idea *before the Big Bang* – without time the phrase is meaningless. The other thing that is wrong with this picture is that if the Universe did consist of a big lump of matter in the middle of empty space, it would be extremely inhomogeneous. Hubble

* The words *Big Bang* were first used by the British astrophysicist Fred Hoyle as a jibe at the theory. Hoyle was one of the creators of the rival, now long-defunct, Steady State Theory. The confusion created by the words is perhaps Hoyle's posthumous revenge.

showed that even today the Universe is fairly homogeneous – whichever galaxy we lived in, we would see roughly the same number of galaxies around us (Chapter 6) – and during the first few hundred thousand years after the Big Bang, according to the theory, it was actually much more homogeneous than it is today.

The Big Bang is, to steal a phrase from Winston Churchill, a riddle wrapped in a mystery inside an enigma. The beginning of the Universe is hidden in a labyrinth of misconceptions, abstract (although simple) mathematics, high-powered theoretical physics (although this is irrelevant for understanding the basic ideas) and philosophical will-o'-the-wisps (the secret is not to follow them). In this chapter and the next, I will try to guide the reader through the labyrinth. Something to hold onto in the middle of the labyrinth will be the thought that there are some close and rather simple connections between life on Earth and the Big Bang.

The first of these connections was uncovered in the 1960s by Arlo Penzias and Robert Wilson, who were engineers working at the Bell Telephone Laboratory in New Jersey, which is about as prosaic a job and location and as far from the romance of astronomy as you can possibly get. They had been set the task of tracking down sources of radio interference – radio noise. They designed and built a special horn-shaped antenna which allowed them to isolate most of the obvious sources of noise. However, after they had accounted for the known sources of noise, they found there was still some noise remaining. The source of this residual noise was a mystery. Its strength was remarkably constant. It was the same during day and night, which immediately eliminated some obvious terrestrial suspects. The constancy of the signal also eliminated the usual astronomical suspects, such as the Sun, since the radiation from one of these is stronger when the object is in the sky. Penzias and Wilson were sufficiently puzzled that they even suspected some pigeon droppings they discovered in the horn. But even after these had been shovelled out and the pigeons scared away, the noise was still there.

The mystery was solved by astronomers just down the road at Princeton University. The Princeton astronomers had recently shown that during the first few hundred thousand years after the Big Bang the Universe was filled with hot gas. The temperature of the gas was approximately the same as the surface of the Sun, and so like the Sun the gas emitted photons of visible light. Because of the expansion of the Universe during the next 14 billion years, the energy of the photons gradually fell – just as a hot gas cools down when it is allowed to expand. The Princeton astronomers calculated that by the present day this radiation would be in the radio waveband. They quickly realized that the mysterious radio noise detected by Penzias and Wilson was exactly the radiation predicted by the Big Bang theory.

One of the connections between events 14 billion years ago and everyday life on Earth is something which is in everyone's living room. The pattern of static seen when a TV is tuned to a frequency between channels is mostly caused by sources of terrestrial noise, but about 1% of the static is caused by the radiation discovered by Penzias and Wilson[16]. It is an understatement to say this connection is remarkable. The radiation detected by the box-shaped radio-telescope in the corner of my living room has been travelling for almost 14 billion years, from a time only a few hundred thousand years after the Big Bang. This radiation fills every part of the Universe today, including the room in which you are reading this book and the room in which I am typing these words (my office, I estimate, contains about one billion photons). This radiation is significant even on the Earth where there are many strong artificial sources of radiation. In the Universe at large, this radiation, left over from the Big Bang, overwhelms every other kind. The energy it contains is about 20 times greater than the energy contained in all the visible light emitted by all the stars in the Universe since the Big Bang[17].

The strength of this radiation is virtually the same in every direction, which was the main argument used by Penzias and Wilson for why the source of the radiation could not be within our Galaxy. During the next two decades, a succession of more and more sophisticated experiments showed that the strength of this radiation is remarkably constant; it varies across the sky by less than one part in ten thousand*. The strength of the radiation in

* There is actually a change in the strength of the radiation from one side of the sky to the other, the result of a Doppler shift caused by the motion of the Local Group (Chapter 6) and by our motion within the Local Group. However, once one allows for this local effect, this statement is correct.

any direction tells us about the temperature and density of gas in that direction 14 billion years ago. Its lack of variation shows that the Universe during its first few hundred thousand years must have been remarkably uniform, with the temperature and density of the gas hardly varying at all from place to place.

However, with a moment's reflection, this is a bit surprising. The world around us contains galaxies, stars, planets, trees, clouds and daffodils. All this diversity however has arisen from a Universe in which there appears to have been almost no variation at all! The Universe today *is* homogeneous, but only on the very largest scale; if I took a big box one hundred million light years on a side and put it down anywhere in the Universe today, I would always find roughly the same number of galaxies in the box. But on a smaller scale than this, on the scale of a galaxy cluster, let alone the scales of stars and planets, the Universe is remarkably lumpy. How did the spectacularly uniform Universe that existed shortly after the Big Bang give rise to the lumpy Universe we see around us today? This paradox was summarized best by the astronomer William Saslaw, who said that "the greatest mystery about galaxies is why they exist at all."

Astronomers have always assumed that the solution to the paradox is a simple one: gravity. Gravity is clearly responsible, after all, for at least the later stages in the formation of the big lumps we see around us: planets, stars and galaxies. Gravity also has the nice property that it does not need much to work with. As long as the Universe shortly after the Big Bang was not completely homogeneous and there were places where the gas density was slightly higher than in others – even if only by a minuscule amount – gravity would have attracted more gas towards these places, slowly increasing the difference in density. Eventually, given 14 billion years, these slight enhancements in density, so astronomers reasoned, would have produced the lumpy Universe we see.

If there were small variations in the gas density 14 billion years ago, these should have given rise to small variations across the sky in the brightness of the radiation we now observe. However, the size that the theorists predicted for these variations was very small, and initially it was not surprising that the observers could not detect them. But as the years wore on, and the radio observations of what came to be called the *cosmic background radiation* became more and more sensitive, this continual failure began to seem rather surprising.

In 1989 NASA launched a satellite designed to settle this issue. The Cosmic Background Explorer (COBE) was actually designed to answer several questions (Chapter 7), but the biggest questions were concerned with the cosmic background radiation*. The instrument on the satellite with the goal of looking for variations in the radiation was the Differential Microwave Radiometer, designed by a team led by George Smoot at the University of Berkeley. The DMR had two big advantages over previous experiments. These had usually been limited by radio noise from the Earth's atmosphere and from other terrestrial sources; COBE's location well above the atmosphere meant that the DMR was immune from these problems. Some previous experiments had been flown on balloons, which had also largely overcome these problems, but an additional difficulty for balloon experiments is the short flight time. Once COBE was in orbit, Smoot's team were able to observe the sky for several years, gradually building up the signal from the cosmic background radiation and allowing the team to make multiple cross-checks on their data.

In 1992 George Smoot called a press conference at the Lawrence Berkeley National Laboratory to announce the team's first results. They had made the first clear detection of variations in the cosmic background radiation. Figure 8.1 is the picture Smoot showed at the press conference, which rapidly appeared on the front pages of newspapers and on the TV news around the world. The picture shows how the brightness of the cosmic background radiation varies across the sky. The variation is only about one part in one hundred thousand (the color scheme has been

* This is not the same as the submillimeter background radiation described in Chapter 7, which was also detected by COBE. The submillimeter background radiation is the combined submillimeter radiation from all the galaxies in the Universe. The cosmic background radiation, which is roughly 40 times stronger, is the radiation from the hot gas that filled the Universe during its first four hundred thousand years.

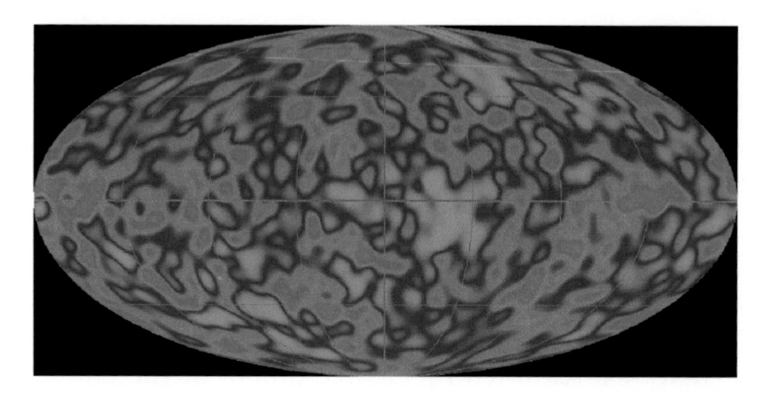

FIGURE 8.1 The image made by COBE of the cosmic background radiation. Darker areas are where the cosmic background radiation is brighter than average, lighter areas are where it is fainter than average.

artfully chosen to accentuate the variation), but it is enough to explain why the Universe today is so lumpy. The brightness peaks in the picture reveal density enhancements, places where 14 billion years ago the density of the gas was slightly higher than average, if only by a tiny amount. Gravity and 14 billion years were enough to transform these tiny density enhancements into the lumpy Universe we see around us.

This is a real picture of the Universe 14 billion years ago – as real as the image of this page in your mind. There are some unimportant differences. When you look at this page, light photons from the page hit the retina at the back of your eye; your brain then transforms the electrical signals from the light-sensitive cells on the retina into a mental image of the page. There is one additional step in the COBE picture, because the DMR scientists have used a computer to represent the radio signals detected by COBE as a visual picture, which the computer in our head then transforms into a mental image. But this picture contains exactly the same information as the original radio signals. The other difference, of course, is in the travel time of the photons. Photons cross the gap from this page to your retina in about one billionth of a second; the photons in the cosmic background radiation have been travelling for 14 billion years.

The method of observing the history of the Universe by looking out into space is a powerful one, but it does have a limit.

Suppose there is something between this page and your eye that scatters light. Some of the photons streaming off the page will no longer travel in a straight line but will be scattered before they reach your eye, mixing up the photons from different words – and if enough photons have been scattered it will no longer be possible to read the words at all. The same thing happens if we look far enough back in time. During its first four hundred thousand years the Universe was hotter than the surface of the Sun. The intense heat stripped the electrons away from atoms, producing an ocean of electrons and atomic nuclei. Electrons are highly efficient at scattering radiation. We can not observe events during this period for the same reason we can not see below the Sun's surface. Photons may reach us from this period, as they reach us from the interior of the Sun, but in both cases they have been scattered so many times that they no longer contain any useful information. Therefore in one respect the spectacular COBE image is as good as we can get (at least with electromagnetic radiation – see Chapter 9): we can look back in time to four hundred thousand years after the Big Bang but no further.

The cosmic background radiation is one of the two pillars of the Big Bang theory. The second pillar of the theory also has a close connection to life on Earth.

Balloons are one of life's minor pleasures – no summer fete or children's birthday party would be complete without a child crying and a balloon disappearing into the distance. The connection to the Big Bang is the gas that fills the balloons. Helium is a rare element on the Earth, but it is very important in the Universe at large. Astronomers have estimated that roughly 78% of the mass of the Universe consists of hydrogen, 20% is helium, and only 2% is the stuff that makes up the Earth and the objects of our everyday world – carbon, oxygen, iron, and so on. In Chapter 4 I described how the elements that make up our world are created by nuclear processes in stars and are then dispersed through the Galaxy by supernovae. Hydrogen is not made in stars, but it is the simplest element, and astronomers assume that it is the primordial stuff out of which the other elements were made. The anomalous element is helium. Of course, helium *is* made in stars – the conversion of hydrogen into helium is the energy source in many stars (Chapter 4) – but it is manufactured in nothing like the quantities required. The best current estimate is that no more than 10% of the helium we see in the Universe today could have been made by nuclear fusion in stars.

The explanation of the anomaly was first suggested by the Russian–American physicist George Gamow back in the 1940s[18]. Gamow realized the Universe was so hot and dense in the first few minutes after the Big Bang that it was effectively a nuclear fusion reactor. He showed that almost all the helium we see today would have formed naturally from hydrogen by nuclear fusion during the first three minutes after the Big Bang – according to Gamow, "in less time than it takes to cook a goose." He also made the spectacular prediction, which was ignored at the time, that if the density and temperature of the gas were high enough to make helium in the necessary quantities, the gas must have emitted a large amount of radiation – radiation that we should now be able to detect with radio-telescopes. When the cosmic background radiation was discovered accidentally twenty years later, and when Gamow's prediction was eventually remembered, the observed brightness of the radiation was only about a factor of two away from his original prediction.

Of course, some people do not believe the Big Bang ever happened. And the statement that there are two pieces of evidence for the Big Bang theory opens the door for them to claim that it should not be taken too seriously because there are *only* two pieces of evidence. However, this would be to misunderstand the nature of the evidence. These pieces of evidence are so compelling because nobody has been able to think of any alternative way of explaining them. There is also a web of smaller pieces of evidence, which together are virtually impossible to explain without a Big Bang, but it is these two results which are the pillars on which the theory rests.

Cosmology is the study of the Universe as a whole rather than the objects in it. The revolution in cosmology that has taken place over the last decade is not that astronomers are now more convinced of the truth of the Big Bang theory than they were ten years ago (although the web of evidence has been getting denser all the time), but that they now have answers to many fundamental questions about the Universe. A good place to start is to consider what we did not know about the Universe in 1995 (ten years ago, as I write). Here are some basic questions for which there were not good answers ten years ago. How old is the Universe? How much matter does it contain? Is space curved? Will the Universe expand for ever or will it eventually collapse? It is not completely true to say there were no answers to any of these questions ten years ago, but the answers that did exist were highly uncertain. Estimates of the age of the Universe, for example, ranged from 6.3 to 19 billion years[19].

Some of these questions are connected. The ultimate fate of the Universe, for example, depends on the amount of matter it contains. A complication in explaining this connection is the cosmological constant, which was invented by Einstein in the early twentieth century (Chapter 6). As I will describe later in this chapter, a pivotal event in the cosmological revolution of the last decade was the discovery that there does, after all, appear to be a cosmological constant. However, for the moment let us return to the simpler world of 1995 when astronomers assumed that the cosmological constant was a figment of Einstein's imagination.

If there is no cosmological constant, it is simple – in principle at least – to discover the ultimate fate of the Universe. We know that the Universe is expanding from the discovery early last century that most galaxies have redshifts (Chapter 6). We also know that because there is a gravitational attraction between every piece of matter in the Universe and every other piece of matter, the expansion must be slowing down. The big question is whether this gravitational drag on the expansion will eventually bring the Universe to a halt. However, if the Universe does eventually stop expanding, gravity will not switch off, and the expansion will become a contraction. In this case, the end of the Universe will be a reverse Big Bang – the *Big Crunch*. The gravity of an object depends on its mass, and some simple theory shows that the strength of the gravitational drag on the expansion depends on the average density of all the matter in the Universe. The average density is the density one would measure if the present lumpy Universe was smoothed out so that it was the same everywhere. According to the theory, there is a *critical density*: if the average density is less than this critical density, the Universe will expand for ever; if the average density is greater than the critical density, the end of the Universe will be the Big Crunch.

In a famous poem Robert Frost described the two possible fates of the Universe as being fire or ice. The fate of our Universe is summed up in the variable Ω_M. This is defined as the average density of the Universe divided by the critical density. If Ω_M is less than one, the average density is less than the critical density, and so the Universe will expand for ever. Stars will continue to form, but eventually there will be no gas left to make stars; and after the last star has died, all that will be left will be a wasteland of burnt-out white dwarfs, neutron stars and black holes – the end of ice. If Ω_M is greater than one, the average density is greater than the critical density, and everything – stars, planets and galaxies – will eventually be consumed in the Big Crunch – the end of fire.

These questions are also connected to the question of the curvature of space. The strange effects of curved space are difficult ones to appreciate because our senses and brains are attuned to life in flat space (if space *is* curved, the curvature is undetectable on the scale of our everyday world). One way of gaining some understanding of these effects is to consider a two-dimensional universe. In Chapter 6 I introduced Fred, a 2D character living on the surface of a sphere, a 2D universe with positive curvature. Although Fred could not see he was living on the surface of a sphere, he would have been able to tell that his universe was curved by carrying out a simple geometric test, such as adding up the angles in a triangle. In the same way, we should be able to tell whether our Universe has positive, negative or zero curvature by carrying out a geometric test, as long as the test is carried out on a large enough scale that the curvature is evident. The theory shows that the curvature of space depends on the value of Ω_M (It also depends on the value of the cosmological constant, but we will return again to the happy days of the middle 1990s when cosmologists did not have to worry about this). If Ω_M is less than one, space has negative curvature; if Ω_M is greater than one, space has positive curvature; and if Ω_M is equal to one, the curvature is zero – space is flat.

The answers to three different questions are therefore contained in the variable Ω_M. By the mid-1990s, a fairly accurate value was known for the critical density – all that was then required, since Ω_M is just the average density divided by the critical density, was an accurate measurement of the average density of the

Universe. Unfortunately, although astronomers had estimated this in various ways, none of the estimates agreed.

In the mid-1990s, therefore, none of the set of basic questions had answers. The cosmological revolution that has since occurred is not the result of a single wonderful project that has told us the answers to all the questions. If that were the case, there would be the worry that there is something nasty lurking in the undergrowth: some problem with the project that nobody has yet thought of. The revolution has actually occurred as the result of a large number of projects – all of which have different methods, assumptions and uncertainties – but all of which give similar answers. Cosmologists now talk about the *concordance model* of the Universe because of the concord between the results of the different projects. In the rest of this chapter I am going to describe three of the more important projects, but it is important to realize that the answers are the result of the work of many different teams of astronomers.

The announcement by George Smoot of the first detection of variations in the cosmic background radiation did not mark the end of a story but rather the beginning. The picture that Smoot showed to the press (Figure 8.1) is only a broad-brush picture of the cosmic background radiation. There is not much detail and, as I will describe later, we now know that a detailed picture of the cosmic background radiation contains one set of answers to all the basic questions. This only gradually became clear during the 1990s, but even at the beginning of that decade it was indisputable that it was important to obtain a more detailed image than the one provided by COBE.

Many teams around the world set out to obtain this image, using a variety of techniques, ranging from interferometers on the ground to telescopes in the stratosphere. Most of these teams were eventually successful, and all their results provided important support for the concordance model. However, as usual, the team which made the biggest splash was the one that got there first.

This team was a group of scientists from around the world led by an Italian, Paolo de Bernardis, and an American, Andrew Lange. Their plan was to observe the cosmic background radiation using a telescope suspended from a balloon. By sending the telescope into the stratosphere, they would be able to overcome most of the problems caused by terrestrial radio noise and by radio noise from the atmosphere itself. The disadvantage of using a balloon, of course, is the wind: often before one has made enough observations the telescope has to be brought down to avoid losing it (and the data) in the ocean. The team planned to overcome this problem using the peculiar wind patterns in Antarctica, which during the summer follow a circular path around the South Pole. They realized that if they launched the balloon at the edge of the continent during the Antarctic summer, the wind would blow the balloon right round the Pole, bringing it back to roughly the same place. This would give them a long enough flight time. In the grand tradition of corny acronyms for astronomy projects, somebody came up with the name BOOMERANG for the telescope, because it would return to the same place and it was just possible, with a bit of creative license, to find suitable words for each letter*.

In autumn 1997, after the usual delay in raising funds, designing and building the telescope and carrying out test flights (about five years in the case of BOOMERANG), the team flew down to McMurdo Scientific Station in Antarctica for the first long-duration balloon flight*. This is a large military-style American base on Ross Island and, as the home of between 200 and 1000 scientists, it is the largest town in Antarctica. Ross Island is the closest solid ground to the Pole that can be reached by ship, and for this reason it was one of the bases of Captain Scott. Scott's original hut is still there, and it is possible to peer through the window at 100-year-old boots and tins of bully-beef on the shelves. The busy time at McMurdo is in the Antarctic summer, when the Sun is in the sky twenty four hours a day and the base is a hive of scientific activity. After the BOOMERANG team arrived at the base, they worked 18-hour days for 12 weeks, checking out every aspect of the telescope. When they were sure (as far as anyone could be sure) that the telescope and its detectors and electronics would work successfully, they handed BOOMERANG over to the

---

* *Balloon Observations of Millimetric Extragalactic Radiation Anisotropy and Geophysics.*

* The chronology and other details of the first long-duration balloon flight that I have given here are based on the memories of Phil Mauskopf, one of the members of the original BOOMERANG team.

National Scientific Balloon Facility, which is responsible for launching the balloons at McMurdo.

Launching a stratospheric balloon is a more difficult art than releasing a child's helium balloon. Stratospheric balloons can not be released at all if the weather conditions are not perfect, and for about five days over Christmas 1998 the BOOMERANG team waited for suitable launch conditions. Finally, on December 29th, 1998, the launch team decided that the wind-speed was low enough and the cloud cover was acceptable. They started preparations for the launch. The astronomers trooped out onto the ice to watch.

The launch site was out of sight of the base, a large expanse of ice and snow with enough space for the mobile launch vehicle (MLV) to manoeuvre, and with Mount Erebus, one of the largest mountains in Antarctica, as a backdrop. While the team watched from a distance, the balloon started to inflate. As it began to rise and the cables tighten, the driver expertly manoeuvred the MLV around on the ice. Just at the perfect moment, when the MLV was directly underneath the rising balloon, the driver released the gondola containing the telescope. The balloon rose silently into the sky. The team stayed to watch as the balloon became smaller and smaller and finally vanished into the distance.

During the next ten days the prevailing winds carried BOOMERANG around the Pole at 130,000 feet. After ten days, measurements with the satellite Global Position System showed the team that BOOMERANG was back in the vicinity of McMurdo. They sent a radio signal which triggered explosive bolts, separating the balloon from the gondola, which plummeted towards the ground. When the gondola reached 50,000 feet, it released parachutes, and the gondola and telescope floated down safely onto the ice.

The first Antarctic launch and retrieval were the two most outwardly dramatic moments in the history of BOOMERANG, but the most important moment occurred a few months later in an office in California. This was when, after the team had completed all the preliminary stages of the data analysis, the first image of the sky appeared on a computer screen. Phil Mauskopf was one of the graduate students in the team and he remembers that his reaction when he saw the first BOOMERANG image was

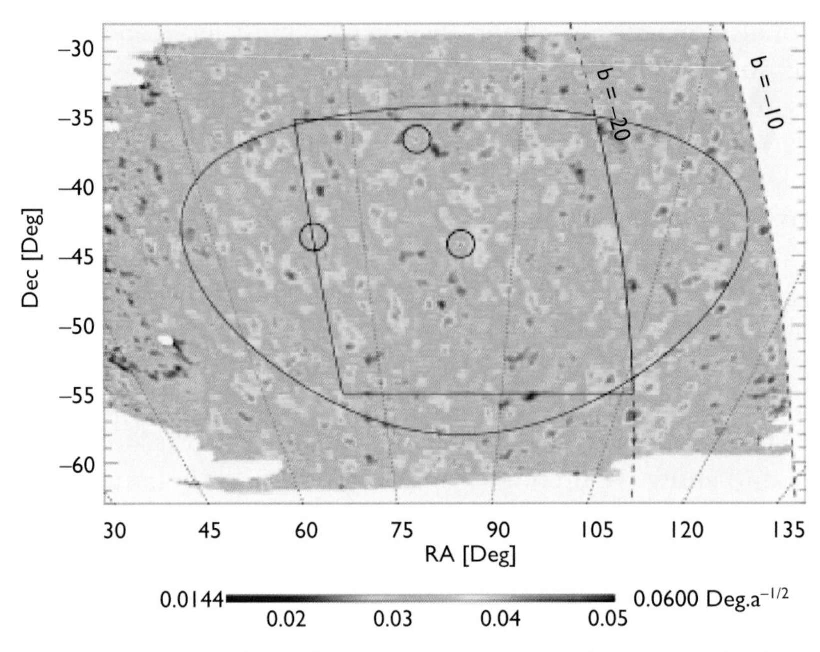

FIGURE 8.2 Image made with BOOMERANG of the cosmic background radiation. The dark areas show where the radiation is brighter than average, the light areas show where it is fainter than average

"Wow!" My reaction when I saw it a year later was "Wow!" Every astronomer with whom I have talked had the same reaction.

Figure 8.2 shows the image. The dark areas are where the cosmic background radiation is brighter than average, the light areas are where it is fainter than average. COBE provided a broad-brush picture of the cosmic background radiation over the whole sky (Figure 8.1). The BOOMERANG image is of only a small part of the sky, but it is a finely drawn picture. The image that appeared on the screen was the first detailed picture of the cosmic background radiation.

Apart from its huge wow factor, the image allowed the team to answer one of the basic questions about the Universe. The best way to start the discussion of how they did this is to think about sound. Any sound – the sound of a saxophone, a child crying or even the noise of a car back-firing – is a combination of sound waves of different frequencies. Scientists can analyze any sound and determine the strength of the contribution at each frequency. The sound from the saxophone is dominated by sound waves with only a narrow frequency range; the sound of the car back-firing contains a much wider range. In a similar way, one can think of an image as being a combination of waves, although in this case it makes more sense to think of the image as being the combination of waves of different wavelength (the wavelength is just the distance between two crests of the wave) rather than waves of different frequency. The two ways of thinking are mathematically equivalent because the wavelength and frequency of a wave are connected by a simple formula.

The BOOMERANG image does look, to me at least, like the combination of many waves. When viewed from above, ocean waves usually appear as a series of parallel ridges and valleys sweeping towards the shore. Admittedly, the BOOMERANG image does not look much like this. But suppose we are looking at a place where several sets of waves are approaching each other. The maelstrom that would be produced might look very much like the BOOMERANG image: high points in the water where the ridges from two sets of waves overlap, dips in the water level where the valleys of two sets overlap. Scientists can mathematically determine the relative strengths of the waves that make up an image in much the same way they can determine the composition of a sound. When the BOOMERANG team applied this mathematical technique, which is called Fourier analysis, to their image, they made the interesting discovery that the image, just like the sound of a saxophone, consists mostly of waves with a small wavelength range (Figure 8.3). The wavelength of these waves is about one degree, which happens to be about twice the angle made in the sky by the Full Moon.

The musical example I started with is particularly appropriate in this case. As I explained above, the ups-and-downs in the BOOMERANG image were caused by variations in the gas density 14 billion years ago. These tiny variations in gas density were actually caused by sound waves propogating through the gas. I will explain later what generated these sound waves. The fact that the waves making up the image have a small range of wavelength shows that these sound waves also had a fairly narrow range of wavelength. This is one place where I have to give proper credit to theorists because a group of theorists had actually predicted this back in the 1960s. The theorists had even been able to calculate

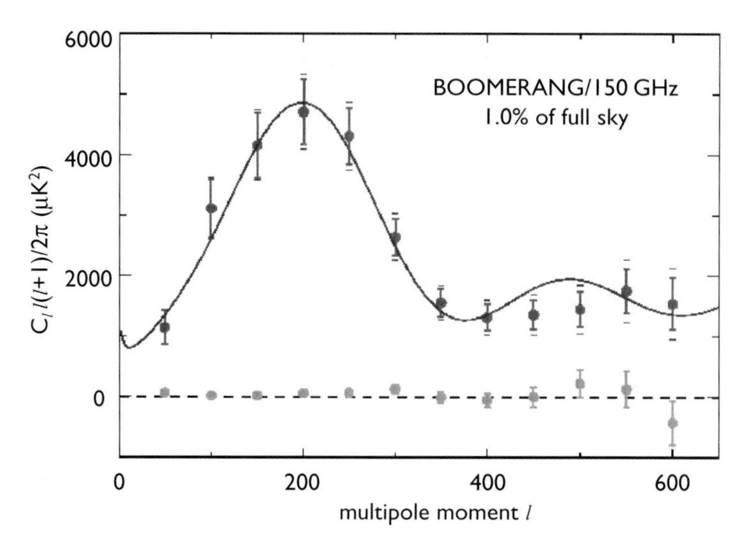

FIGURE 8.3 The BOOMERANG team's decomposition of their image into different waves. The actual things plotted are specialized mathematical terms, but effectively the wavelength of the waves is plotted along the horizontal axis and the strength of the waves is plotted along the vertical axis. The peak in the curve shows that the image is dominated by waves with a narrow range of wavelength.

the typical wavelength of these sound waves in light years—600,000 light years. The BOOMERANG team were able to use the theorists' calculation and their own measurement of the angle the waves make on the sky to answer one of the set of basic cosmological questions.

All one needs to do to measure the curvature of space is to carry out a geometric test, as long as the test is carried out over a large enough region of space for the curvature to be evident. The example I used earlier was adding up the angles in a triangle, and the BOOMERANG team had at their disposal the largest possible triangle. One side of the triangle was the distance (over ten billion light years) the photons had travelled to reach their telescope. A second side of the triangle was the characteristic wavelength of the sound waves in light years, which had been calculated by the theorists in the 1960s. The BOOMERANG team also knew one of the angles in the triangle, the angle these waves make on the sky, which they had measured themselves. With two sides and an angle, it's possible to finish drawing the triangle, and for geometric tests measurements of two sides and one angle are as good as measurements of three angles. The BOOMERANG team were able to use their results to make a simple test of the curvature of space.

Their answer to the question of whether the Universe has positive, negative or zero curvature was that the Universe as a whole has zero curvature – on the largest possible scale, space is flat. They announced their answer in a letter to the journal *Nature* in April 2000.

As I described earlier, some of the fundamental questions about the Universe are connected; and if in the year 2000 the cosmological assumptions of the mid-1990s had still been unquestioned, the BOOMERANG team's measurement would also have provided answers to two other questions. However, as the result of another project, the standard cosmological ideas of over four decades had already flown out of the window.

The second project I want to describe was a more conventional astronomy project than BOOMERANG. The Supernova Cosmology Project was carried out by an international team of astronomers led by the American astronomer Saul Perlmutter. Perlmutter's team had resurrected Hubble's old idea of standard candles (Chapter 6). The idea, remember, was to find a type of galaxy with a narrow range of luminosity and then plot how the brightness of these *standard candles* varies with distance. In our everyday world, brightness varies with distance in a particular way, but the relationship in an expanding Universe will be different. The relationship will actually depend on the value of the variable Ω_M, which is defined, as I explained above, as the average density of the Universe divided by the critical density. Therefore, in principle, by using the standard candle technique, it is possible to measure Ω_M and thus also answer the question of the ultimate fate of the Universe. Astronomers however had largely discarded this technique in the late 1970s, because they eventually realized that galaxies are never standard candles; galaxies evolve, and so a galaxy far enough out in space for the method to be useful is being seen so far back in time that its luminosity is probably different from galaxies of the same kind today (Chapter 7).

In the late 1980s, however, astronomers realized that they could still use the standard-candle method if, instead of using galaxies, they used supernovae. In Chapter 4 I described our current ideas about the cause of a Type 1a supernova, but the truth is: cosmologists don't care. As long as all the objects in a class have only a small range of luminosity, they can be used as standard candles, and a cosmologist does not need to know anything more about them. Obviously the brightness of a supernova changes dramatically in a very short time, but at the moment of maximum light the luminosity of all supernovae in the Type 1a class appears to be very similar. The big advantage of supernovae over galaxies as standard candles is that there is no reason to expect that a Type 1a supernova that occurred five billion years ago should have been any different from a Type 1a supernova that occurs in the Universe today.

Of course, it is not quite that simple in practice – otherwise astronomers would have used supernovae as standard candles many years earlier. The main problem is that supernovae are rare: a supernova only occurs in a galaxy about once every 30 years. Both Perlmutter's group and a rival team, the High-Redshift Supernova Search Team, spent over a decade monitoring tens of thousands of galaxies in order to find enough supernovae to apply the method. Even after the teams had discovered a supernova, they still had to make many additional observations before the supernova could be used as a standard candle. They had to monitor the brightness of the supernova to find the moment of maximum light; they had to measure the redshift of the galaxy in which the supernova had occurred; and they also had to obtain a spectrum of the supernova itself to be sure that it was not one of the other types of supernova. All this required years of work and the use of many different telescopes, including the Keck Telescope and the Hubble Space Telescope. Nonetheless, by the late 1990s, Perlmutter's team had obtained a sample of supernovae that was large enough to be useful.

They expected that by plotting the brightness at maximum light of each supernova against the redshift of the galaxy in which it had occurred, they would be able to estimate the value of Ω_M. If gravity is the only important long-range force, everything about the Universe – its future history, its current age, the relationship between redshift and time (Chapter 7) *and also the relationship between brightness and redshift for standard candles* – depends on its average density and thus on the value of Ω_M. As long as Type 1a supernovae *are* standard candles, the theory shows that at a given redshift, the brightness of a supernova will depend only on the value of Ω_M; the higher the value of Ω_M, the brighter the supernova will appear.

However, when the team plotted this diagram, they received a shock. The supernovae at the highest redshifts were fainter than was predicted by the theory, *whatever value of Ω_M was used.* According to the theory, supernovae would be faintest for the lowest possible value of Ω_M, which is zero. This is obviously an unrealistic value, because it implies that the average density of the Universe is zero, but even with this value the team found the supernovae were fainter than the theory predicted.

When theory and observations are in disagreement, there must be something wrong with either the theory or the observations. As good observers, Perlmutter's team decided the theory must be wrong. The simplest way they could find to fix the discrepancy was to assume that the cosmological constant is not zero. The cosmological constant, remember, was something dreamt up by Einstein in the 1920s because of a mistake (Chapter 6). Einstein had realized that the equations of general relativity imply that the Universe is dynamic, that it must be either expanding or contracting – and since he wrongly thought that it was static, he added an extra term, the cosmological constant, to his equations to make the Universe stand still. When he belatedly realized the significance of the galaxy redshifts, he admitted his blunder, and scientists consigned the cosmological constant to the dustbin of history. However, a cosmological constant does give observers an extra degree of freedom. If the cosmological constant was not just a figment of Einstein's imagination, everything about the Universe – its future history, its current age, the relationship between time and redshift, and so on – depends on the value of Ω_M *and* also on a variable Ω_Λ, which is effectively the cosmological constant. The brightness that the theory predicts for a high-redshift standard candle is less if the cosmological constant is not zero, and Perlmutter's team found that with a suitable value for the cosmological constant the predicted and observed brightness of the high-redshift supernovae came nicely into agreement. Perlmutter's team announced their discovery that there is, after all, a cosmological constant in a paper in the *Astrophysical Journal* in 1999. The other supernova team obtained a similar result. The results of the two teams implied that an assumption that astronomers had been making about the Universe for eight decades was wrong.

I thought *they* were wrong. I heard about the result of Perlmutter's team in the way I hear about most important astronomical discoveries: listening to the BBC news while washing up the breakfast dishes. My reaction at the time, I remember, was scepticism. Even with hindsight, this seems a perfectly sensible reaction. It is easy enough to think of things that might cause the result to be wrong. For example, although we have no reason to suspect that Type 1a supernovae are not standard candles, our understanding of why they occur at all is so poor that it is perfectly possible there is some unknown effect that causes a high-redshift supernova, which after all happened billions of years in the past, to be less luminous than a supernova today.

The event which made me and, I suspect, most other cosmologists take this result more seriously was the announcement a year later of the discovery of the BOOMERANG team that space is flat. The reason for our greater confidence was that the BOOMERANG result in combination with an earlier cosmological result provided an independent check on the results of the supernova teams.

To explain why this was, I need to take a brief detour to describe two of the ways in which astronomers have tried to measure the average density of the Universe, which is needed to estimate the value of the important variable Ω_M. The simplest way of measuring this is bean-counting (as it is often called in the trade). Draw a large box in space, large enough that the box contains a fair sample of the Universe. Then estimate the mass of each galaxy in the box from the amount of light it produces, by making the assumption that all the light is produced by average stars like the Sun. A very luminous galaxy, for example, might have a luminosity that is one thousand billion times the luminosity of the Sun; its mass must therefore be, if the assumption is correct, one thousand billion times the mass of the Sun. Once the masses of the individual galaxies are known, add up the masses of all the galaxies in the box. Then divide the total mass by the volume of the box – and this gives you the average density. Using this very simple method, astronomers have estimated that the value of Ω_M is about 0.004.

There are however some obvious problems with this method. The most fundamental one is the assumption that all the matter in the Universe is in stars, which is obviously wrong because of the existence of interstellar gas and dust (Chapter 5). If there is nonluminous material within a galaxy or even material between the galaxies, this method will yield too low a value for the average density. Fortunately, there is a simple way to check this assumption, one that in principle should yield a much more accurate value for Ω_M.

The most accurate way of estimating the mass of the Solar System is to use Newton's law of gravity: since the speed of a planet depends on the gravitational force of all the matter inside its orbit, the speed of the outermost planet will yield a good estimate for the total mass. Astronomers have estimated the masses of galaxies in much the same way. By measuring the speed at which either stars or gas clouds orbit the center of a galaxy, they have shown that galaxies contain at least as much *dark matter*, matter that does not emit light, as is contained in the stars. It is a bit trickier to apply Newton's law to a still larger object, the Universe itself, because in the Universe there is no *outermost galaxy*, merely many galaxies moving about in different directions and with different speeds. Another complication is that as one looks further out into the Universe, the speeds of galaxies relative to our own increase because of the expansion of the Universe. However, using a variety of different techniques, which all boil down to using Newton's law, astronomers have tried to estimate the average density of the Universe. By the mid-1990s, they were typically finding values for Ω_M of between 0.1 and 0.3.

This was a surprising result because it implies, when one compares it with the value obtained by bean-counting, that there is at least 20 times more dark matter in the Universe than is contained in stars. Partly because it was such a surprise and partly because during the 1990s most theorists preferred (for reasons I have not got time to go into) a value for Ω_M of 1.0, astronomers at the time did not know quite what to make of this result.

However, now let us move forward in time by five years and consider this result together with the result of the BOOMERANG team. We will put aside for the moment the claim of the supernova teams that there is a nonzero cosmological constant. This is the one place in this book where I am forced to write down an equation (although a very simple one). The BOOMERANG project showed that space is flat, and some simple theory shows that if the curvature of the Universe is zero the following equation is true:

$$\Omega_M + \Omega_\Lambda = 1.$$

Ω_Λ, remember, is effectively the cosmological constant. The observers in the 1990s had estimated that the value of Ω_M falls in the range 0.1–0.3. If we substitute these values for Ω_M into the equation in turn, we find that the value of Ω_Λ falls in the range 0.7–0.9. This is without considering the results of the supernova teams at all. Thus the combination of the earlier measurements of Ω_M with the BOOMERANG result shows that, *even if the Supernova Cosmology Project had never happened*, there must be a cosmological constant. The predicted value is also about right, because Perlmutter's team found that to bring the predictions for the brightness of the high-redshift supernovae in line with the observations they needed a cosmological constant with a value of about 0.7.

Thus just after the millennium, there was for the first time a fragile consensus between the results of different cosmological projects. The project that has strengthened this consensus is the final one I want to describe in this chapter.

In 2003, the Wilkinson Microwave Anisotropy Probe provided another picture of the cosmic background radiation (Figure 8.4). WMAP was a space telescope like COBE but with the advantage of ten years development in technology. COBE provided a broad-brush picture of the cosmic background radiation over the whole sky; BOOMERANG provided a detailed picture over a small part of the sky; WMAP provided a picture over the whole sky again but drawn with the same fine detail as the BOOMERANG picture. A comparison of this picture with the COBE picture in Figure 8.1 shows how far this art had progressed in a decade.

The WMAP team used the mathematical technique of Fourier analysis to calculate the strengths of the different waves making up the image. As I described earlier, this analysis also reveals the strength of the sound waves of different wavelengths that were travelling through the Universe during the first few hundred thou-

FIGURE 8.4 Image made by the Wilkinson Microwave Anisotropy Probe of the cosmic background radiation. Lighter areas are where the radiation is slightly brighter than average, darker areas where the radiation is slightly fainter than average. Credit: NASA/WMAP science team

sand years after the Big Bang. Figure 8.5 shows the WMAP results. The big peak shows, as the BOOMERANG team had discovered three years before, that the strongest sound waves have a narrow wavelength range – the music contains a dominant note. However, as the figure shows, there are also other smaller peaks – harmonies on the main note. This cosmic music contains a complete set of answers to many of the fundamental questions about the Universe.

During most of the history of the Universe the dominant historical process has been gravity. Gravity is such an overwhelming force that star-formation theorists, for example, do not have a problem explaining the existence of stars, but rather why all the gas in the Galaxy does not immediately collapse to form stars (Chapter 5). During the first 400,000 years, however, the gas filling the Universe was so hot that its pressure was high enough to resist gravitational collapse. This gas was a mixture of atomic nuclei, electrons, and photons of radiation. This mixture is called by cosmologists the *photon–baryon fluid* because atomic nuclei are made up of baryons: protons and neutrons. The sound of a saxophone is produced by a sudden oscillation in air pressure within the instrument, which travels in a wave through the air. Sound waves also travelled through the photon–baryon fluid. Although the frequency of one of these sound waves was very different to the frequencies contained in the wail of a saxophone, the basic

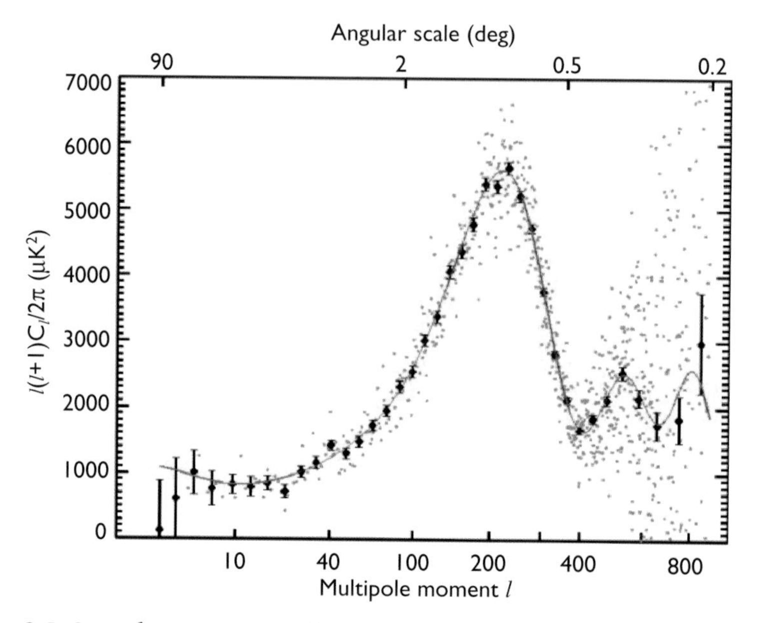

FIGURE 8.5 Sound waves travelling through the Universe four hundred thousand years after the Big Bang. The strength of the different waves is plotted along the vertical axis. The variable plotted along the horizontal axis is effectively the frequency of the sound wave, with the frequency increasing from left to right.

physics of the sound waves was the same: an oscillation in pressure which travels through a gas.

Something is always needed to start a sound wave. In a saxophone, the sound waves are produced by a vibrating reed which the musician blows into. The source of the cosmic music was probably a process during the period of inflation, which occurred during the first split second after the Big Bang, and which I will describe in the next chapter. The harmonies in Figure 8.5, however, were created during the next 400,000 years.

The best way to understand how this happened is to think about the saxophone. In a saxophone, sound waves generated by the vibrating reed travel through the instrument to its horn. The sound waves that produce the greatest effect are the ones that oscillate a whole number of times – 1, 2, 3 etc. – as they travel from the reed to the horn because these produce the greatest change in air pressure at the horn. Sound waves with frequencies that mean they oscillate a fractional number of times as they travel from the reed to the horn (3.17 times, for example) produce a much smaller change in air pressure. These special frequencies are the natural frequencies of the saxophone and are what gives it its distinctive wail. In the Universe, the sound waves were generated at the time of inflation (the reed) and then travelled through the Universe for 400,000 years until their effect was recorded on the cosmic microwave background (the horn). The sound waves that oscillated a whole number of times in this period were the ones that produced the greatest effect on the COBE, BOOMERANG, and WMAP images. The frequencies of these sound waves are the natural frequencies of the Universe.

Cosmologists have been able to use the peaks and valleys in Figure 8.5 (the technical term used by cosmologists is "wiggles") to provide independent answers to the set of basic questions. They have been able to do this because, although gravity was not the overwhelming historical process that it was later on, it was still important during the first 400,000 years. At some places and times gravity was in competition with pressure; at other places and times the effects of the two reinforced each other. Suppose that a sound wave travelled through a region of space where the density was slightly higher than average, and so the gas was starting to contract because of the increased gravitational force. If, as the result of the sound wave, the gas pressure at that moment was at a maximum, the pressure would have resisted the gravitational contraction; if it was at a minimum, the two would have acted in concert. The effect of gravity depends on the amount of matter in the Universe. The theorists in the WMAP team found they could only reproduce the positions of the second and third peaks in Figure 8.5 with a value for the average density of the Universe that gave a value for Ω_M of close to 0.27[20].

They also discovered something more surprising. The photon–baryon fluid experiences the effects of both pressure and gravity. But the WMAP team found they could only reproduce the heights of the wiggles in Figure 8.5 if most of the matter in the Universe only experienced the effect of gravity during this period. This matter must therefore be completely unlike the protons and neutrons, the baryons, which make up our everyday world. This matter has been named *cold dark matter* and its existence had been suspected before, but the WMAP wiggles have provided the clinching evidence that it really exists. During the first 400,000 years, any variations in the density of the cold dark matter continued to grow under the influence of gravity, completely unaffected by the sound waves in the photon–baryon fluid. The big surprise was the amount of cold dark matter. To produce the shape of the wiggles, the WMAP team found that only 16 per cent of the matter in the Universe is baryons; everything else is cold dark matter.

The value of Ω_M derived by the WMAP team was nicely in agreement with the value obtained a decade earlier from simply applying Newton's law to the motions of galaxies. Using the figure of 16 percent, the WMAP team were able to estimate a value for Ω_M from the baryons alone of 0.044. This too was in agreement with an earlier estimate, which I have not had space to describe. The WMAP team were also able to estimate a value for Hubble's constant (Chapter 6), and this too agreed with estimates made in other ways. Thus the WMAP results considerably strengthened the consensus among cosmologists that we now have answers to our set of basic questions. Finally, the WMAP team also used the wiggles to determine an accurate age for the Universe for the first time: 13.7 billion years with an uncertainty of only about 0.4 billion years.

Thus we now have answers to the set of basic questions. We know fairly accurately the average density of normal matter in the Universe. We also know that normal matter, the stuff that makes up our everyday world, only makes up about 16 per cent of all matter. We know that space is flat. We know the age of the Universe. We also know the answer to a question that was not even being asked a decade ago: the value of the cosmological constant. We have reliable answers to all these questions, because for each question we have at least two consistent answers from independent methods.

We also know the ultimate fate of the Universe. Even without the cosmological constant, the Universe contains too little matter for its expansion to ever come to a stop. The effect of the cosmological constant will be to accelerate this expansion. The Big Crunch will never happen and, in Robert Frost's words, the end of the Universe will be ice.

But I want to finish this chapter with questions rather than answers. Answers to this set of questions have been the goal of cosmologists for over seven decades, but the answers have thrown up two new questions. We do not know the composition of the cold dark matter, although particle physicists have some speculative ideas. We have even less understanding of the cosmological constant, which was after all an inelegant term added by Einstein to his equations for spurious reasons – in scientific jargon, a kludge. Astronomers have started calling the mysterious phenomenon which is making the Universe expand faster and faster, *dark energy*. But naming it doesn't explain it.

Answers on a postcard, please.

9. Plato's Ghost

Every year, at the beginning of a lecture course I teach at Cardiff University on cosmology, I ask the students to come up with a list of all the questions they have ever had about life, the Universe and everything. With some prompting from me, they usually come up with a list of between ten and twenty (Figure 9.1 shows the list from earlier this year with a few additions from me). Just as the class is getting excited by a lecture course that for once is going to address all the big issues of life, I dash their hopes by ticking the one question my course on the Big Bang theory is actually going to answer. The point of this exercise is to show both the power and limitations of the theory. The theory is powerful because it only tries to address a narrow range of questions. However, a philosopher from an earlier more ambitious era, such as Plato who invented his own cosmological theory, would never have been satisfied with it. In the second half of this chapter, in deference to Plato, I am going to discuss whether it is possible to broaden the scope of our modern scientific theory of cosmology to address more of these big questions. During the first half of this chapter, however, I am going to continue the observational slant of this book and consider what nitty-gritty facts we can possibly discover about the first moments after the Big Bang.

The Big Bang theory can be used to predict what will happen to the Universe in the future (Chapter 8), but also to discover what has happened in the past. A good example of how scientific laws can be used for both prediction and history is the flight of a ball. Let us suppose that it is a hot summer's day and you are either lazing on the grass watching a game of cricket or are munching on a hotdog watching a game of baseball. You see a ball flying through the sky. You happen to have brought to the game a case of scientific instruments (you are clearly a workaholic who finds it

1) What is the meaning of life?
2) Does God Exist?
3) How did the Universe begin?
4) How will the Universe end?
5) How can we be sure that everything around us is real and not an illusion generated by our mind? (the Matrix question, for those that have seen the movie)
6) I understand consciousness from the inside (I assume you do too, if you are not a figment of my imagination), but is it possible to define 'consciousness'?
7) Why should we be good?
8) What do we mean by 'beauty' and 'truth'? Are they, as Keats claimed, the same thing?
9) Why does space have three dimensions?
10) Why does one plus one equal two? Putting this more formally, why can we describe the Universe using a system of marks on paper that we call 'mathematics'?
11) Why is the Universe comprehensible at all?
12) What do we mean by love?
13) Since I am able to get out of bed in the morning and stumble downstairs to get ready for work, I have a good practical knowledge of space and time? But is it possible to define these concepts in a more objective way?
14) The speed of light, the mass of the electron, the relative strengths of the strong nuclear and gravitational force, and all the other constants of nature have particular values. Why do they have these values?

FIGURE 9.1 Life's Big Questions.

impossible to relax). You use these to measure the speed, position and direction of the ball. With these measurements, you can now use Newton's law of gravity and his laws of motion to calculate the future trajectory of the ball (admittedly, it would be rather easier just to watch where it lands). You can also use the same laws to calculate what the past trajectory of the ball must have been for it to have passed you at that instant with that speed and direction. In exactly the same way, astronomers can use the equations of the Big Bang theory and their measurements of the present properties of the Universe – its current density, its rate of expansion and the value of the cosmological constant – both to forecast its future and to uncover its past.

Table 9.1 shows the results of applying these equations for various times in the past. The table shows the size of the Universe at each time as a fraction of its size today and also its average

TABLE 9.1 Important dates in the history of the Universe

Age of Universe	What was happening	Temperature (in Kelvin*)	Average energy of a subatomic particle (in electron volts)	Size of Universe as a fraction of its size today
0	Big Bang	Infinite	Infinite	Zero
10^{-43} seconds	Planck time	10^{31}	10^{27}	10^{-31}
10^{-36} seconds	Strong nuclear, weak nuclear and electromagnetic forces can not be distinguished before this time.	10^{28}	10^{24}	10^{-28}
10^{-10} seconds	Weak nuclear and electromagnetic forces can not be distinguished before this time	10^{15}	10^{11}	10^{-15}
2 seconds–3 minutes	Period in which helium was formed.	10^{10}–10^{9}	10^{6}–10^{5}	10^{-10}–10^{-9}
400,000 years	Time at which the cosmic background radiation was emitted. This is the end of our ability to observe history (Chapter 8).	2900	0.2	10^{-3}

* The temperature in Centigrade is equal to the temperature in Kelvin minus 273.

temperature. I have used the standard mathematical shorthand for very large and very small numbers. The number 10^4 is 10,000, a one followed by four zeros; the number 10^{-4} is 0.0001, the superscript showing that the one is in the fourth position after the decimal point. The table shows that the shorter the time is since the Big Bang, the higher the temperature of the Universe and the smaller its size.

These equations however do not allow us to say anything meaningful about the Big Bang itself. The Big Bang was a *singularity*. This is a technical mathematical term, but it is easy enough to understand. Suppose we use the equations to try to approach the beast at the center of the labyrinth – the moment of the Big Bang. The equations show that both the temperature and size of the Universe depend on the square root of the time since the Big Bang. The square root of *X* is the number which when multiplied by itself gives *X*. The square root of four is therefore two. Suppose we used the equations to calculate the temperature and size at a particular time – let us say one second after the Big Bang. Let us now reduce this time by a factor of four. It is now 0.25 seconds after the Big Bang and, according to the equations, the temperature must be higher by a factor of two (the size of the Universe must have decreased by the same factor). Let us now decrease the time by another factor of four. It is now only 0.0625 seconds after the Big Bang and the temperature is higher by another factor of two. Let us decrease the time by another factor of four – it is now 0.015625 seconds after the Big Bang and the temperature is higher by another factor of two. We can keep on doing this for ever. Every time we step closer and closer to the beast, but we never quite get there. I could name a minuscule amount of time since the Big Bang, but you could always think of an even smaller interval of time – and the equations would show that the temperature of the Universe was higher and its size smaller than at my time. The alternative to approaching the beast in this step-by-step fashion would be simply to insert into the equations a value for time of zero. If we do this, the equations reveal that at the moment of the Big Bang, the Universe's size was zero and its temperature and density were infinite. The concepts of a universe with zero size and of infinite temperature and density have little meaning for creatures like us, finite in both space and time, and so this is not much better as a way of visualizing the beast. We can use the equations to approach the beast, but ultimately we can never see it.

There are some other limitations on our ability to use these equations to uncover the past. The most fundamental problem is that the equations may tell us the temperature and density of the Universe at different times, but they do not tell us about how matter behaves at these temperatures and densities. Of course, we know from countless physics and chemistry experiments how matter behaves at the temperatures and densities of our everyday world, and we can use this knowledge to travel a surprisingly long way back in time before this problem becomes important.

Let us now, in our mind's eye, start travelling back in time towards the Big Bang. The first important date we come to (Table

9.1) is 400,000 years after the Big Bang. This is a crucial date because it marks the end of our ability to observe history. Before this time the Universe was ionized, which obscures our view in the same way that the center of the Sun, a ball of ionized gas, is hidden from our view (Chapter 8).

The next important dates are two seconds and three minutes after the Big Bang. Most of the helium we see around us today was formed between these two times (Chapter 8). The temperatures and densities are still within the range in which the behavior of matter is understood by physicists. We cannot observe events during this period, but in the same way that we think we understand the processes in the center of the Sun because nobody has been able to think of any other way of explaining the Sun's exterior properties, cosmologists are confident that they understand the Universe at this time because nobody has been able to think of any alternative way of explaining the amount of helium we see around us today.

The next important date as we travel back in time is the ridiculously early time of 10^{-10} seconds after the Big Bang. The temperature at this time was approximately one hundred million times the temperature in the center of the Sun. Nevertheless, cosmologists are still fairly confident that they understand the behavior of matter even at this high temperature, because there are at least two places today where the energies of the particles in the Universe at this time have been reached. One is under the border between Switzerland and France and one is under a field a few miles from Chicago.

The world's most powerful particle accelerators are at Fermilab near Chicago and at CERN on the French – Swiss border. At both institutions, physicists send subatomic particles hurtling through tunnels deep under the ground, and when their speed is high enough the particles are allowed to collide. The crème de la crème of particle accelerators will be the Large Hadron Collider, which is being constructed at CERN and will be opened for business in 2007. The LHC consists of an evacuated circular tunnel 27 km in diameter. Two beams of protons will travel around this circular track in opposite directions, repeatedly crossing and recrossing the border between France and Switzerland, all the time getting faster and faster. By the time the CERN physicists flip the points and allow the trains to collide, the energy of each proton will be approximately 10^{13} electron-volts*, which is about one hundred times higher than the average energy of particles 10^{-10} seconds after the Big Bang. Even before the advent of the LHC, the energies reached in particle accelerators are as high as the energy of the particles in the Universe at this time, and so cosmologists are fairly confident they understand the behavior of matter even at such a short time after the Big Bang.

The language of particle physicists is a much more difficult one than most scientific languages, because its lexicon is one part English to two parts math, and its grammar is drawn from the strange world of quantum mechanics. Nonetheless, astronomers have been forced to learn this language (or at least how to communicate with particle physicists in grunts and sign language), because the experiments at CERN, Fermilab and other laboratories have begun to reveal how matter behaves at the high temperatures and densities that existed shortly after the Big Bang. Scientists at CERN, for example, discovered the crucial evidence that this date, 10^{-10} seconds after the Big Bang, was one of the most important dates in the history of the Universe.

In the Universe today there are four forces of nature: gravity, the electromagnetic force, the weak nuclear force and the strong nuclear force. Gravity needs no introduction. The electromagnetic force sometimes manifests itself as electricity and sometimes as magnetism, and without it virtually everything in my house, including this computer, would stop working. The other two forces however do not usually intrude on our notice, because they are short-range forces which only operate inside the atom. According to the quantum view of nature, a force is not a metaphysical phenomenon operating across empty space but is produced by the exchange of particles. The analogy which is often used is of two people throwing balls to each other; each time one of them catches a ball he feels a small force pushing him away from the other person (if this is not obvious, imagine doing it with rocks). This analogy however is a bit confusing because the exchange of balls can only produce a repulsive force, whereas in the strange

* The unit of energy used by particle physicists is the electron-volt.

quantum world the exchange of particles can also produce an attractive force. In the quantum view, the electrical attraction between a proton and an electron is produced by the exchange of photons. These are however not photons one can actually see. From the quantum perspective, one should think of an electromagnetic *field* permeating the whole of space and constituted by *virtual photons*, ghostly photons that flicker in and out of existence. During their transitory existence, these photons may interact with particles, and it is these interactions that ultimately produce the image on my laptop. (If this all sounds faintly unbelievable, see my excuse in Chapter 4.)

In the 1960s, three scientists, the Pakistani physicist Abdus Salam and the American physicists Sheldon Glashow and Steven Weinberg, invented a theory that explained the weak nuclear force. The essence of the theory is that the weak nuclear force is produced by the exchange of particles with a high mass. In contrast, the particle that mediates the electromagnetic force, the photon, has zero mass, and the result of the mass difference is a difference in the range of the two forces. If I resort again to the analogy of throwing balls, the weak nuclear force has a short range because the balls have a high mass and so cannot be thrown very far and the electromagnetic force has a long range because the balls have no mass at all. However, an important prediction of the theory was that in reality the electromagnetic and the weak nuclear force are actually two aspects of a more fundamental *electroweak* force. According to the theory, we only see the two aspects of this single force when the energies of particles are low, as they are in the Universe today. At times earlier than 10^{-10} seconds, however, the energies of particles were so high that the two forces were indistinguishable and there was a single electroweak force. In 1983 scientists at CERN discovered the two particles predicted by the theory which mediate the weak nuclear force: the W and Z bosons.

Thus 10^{-10} seconds after the Big Bang is a date to remember, because before this date the four forces of nature were actually three.

The next important date is 10^{-36} seconds after the Big Bang. There is a set of theories, usually called the Grand Unified Theories (GUTs), which claim that before this time there were only two forces of nature: gravity and a single force that unified the electroweak and the strong nuclear force. The sign that we have just entered the realm of speculation is that, rather than a single theory, there are, according to one count, at least fourteen different GUTs. Moreover, as the average energy of particles in the Universe at this time was roughly one hundred billion times greater than the energy of the particles even in the Large Hadron Collider, our ability to decide which GUT is correct is very limited.

The last important date (apart from the moment of the Big Bang itself) was 10^{-43} seconds after the Big Bang. The odd one out of the four forces is gravity. Gravity is the only force that has not been reconciled with the quantum view of the world. Physicists have succeeded in explaining three forces as the result of interactions between subatomic particles, but in the general theory of relativity gravity is explained as the consequence of the curvature of space (Chapter 6). The Holy Grail for theoretical physicists such as Steven Hawking has been a theory of *quantum gravity* which would explain gravity in a similar way to the other forces. It would not be true, however, to say that the need for a theory of quantum gravity is very urgent. On the scale of atoms, gravity is insignificant compared with the other forces, and it only becomes a significant force on the scale of planets. But on this scale quantum effects are unimportant and general relativity works perfectly well. We can muddle along well enough with the general theory of relativity in virtually every situation, and physicists have come across only a few places where a theory of quantum gravity is actually needed. One of these is the so-called *Planck time*, 10^{-43} seconds after the Big Bang, when the Universe was so tiny that quantum effects must have been important.

The account that I have given in the last few pages of the early history of the Universe is very unsettling for any observer, because it is based on hardly any observations. We lose our ability to observe history 400,000 years after the Big Bang, and before that time the history is mostly based on theory. As an observer, I am happy enough with this history back to two seconds after the Big Bang because of one overwhelming piece of evidence. The Universe must have been pretty much as the theorists believe between two seconds and three minutes after the Big Bang to explain the amount of helium we see in the Universe today. But are there any observations that tell us anything about the Universe before this time?

There are two and they are properties of the cosmic background radiation. A comparison with the Sun may be helpful here. We cannot see below the photosphere of the Sun, effectively its surface, because the electrons in the Sun scatter radiation. However, astronomers have discovered that the surface is oscillating slightly, and these oscillations contain information about what is happening deep inside the Sun. In the same way, we cannot observe the Universe directly before 400,000 years after the Big Bang, but the cosmic background radiation contains information about its earlier history.

The subject of the last chapter was the tiny variations in the cosmic background radiation. These variations were caused by sound waves travelling through the gas which filled the Universe during its first few hundred thousand years. As I explained at the end of that chapter, the harmonies in the cosmic music we observe in the cosmic background radiation can be explained if these sound waves were generated immediately after the Big Bang and then travelled through the Universe for the next 400,000 years – a voyage which modulated the original frequencies. Astronomers can fairly easily calculate the original frequencies and strengths of the sound waves. Some process very early in the history of the Universe must have produced sound waves with just these strengths and frequencies. This is the first clue.

The second clue is the same thing that first allowed Penzias and Wilson to recognize the importance of this radiation: its remarkable uniformity. The variations in brightness in the pictures in the last chapter look spectacular, but that is because the colors have been chosen to dramatize the differences; the true variation is very small, about one part in one hundred thousand. Apart from these minuscule differences, the brightness of the cosmic background radiation is remarkably constant. It is the same on one side of the sky as the other*. This is actually quite surprising.

---

* There is actually a change of about one percent in the brightness of the cosmic background radiation from one side of the sky to the other, the result of a Doppler shift caused by the motion of the Local Group of galaxies and by our motion within the Local Group. But once a correction is made for this effect of our cosmic neighborhood, this statement is true.

Our world is a web of causal connections. If I touch this key, the letter k appears on the screen; the sound of a car causes my dog to bark, which causes me to look up; a child outside the window hits a ball into the sky, which is itself a long sequence of causal connections – the child's brain sends an electrical signal to a muscle, the muscle contracts, the bat hits the ball. The only limit on this web of connections which surrounds all of us is the speed of light. I could bring someone in China into my causal web by dialling a random Chinese number and muttering something in English down the phone, but I would not be affecting this person *now*; I would be doing it a split second in the future. The ultimate limit to this causal web is set by the age of the Universe. The Universe is 13.7 billion years old. Therefore, even if the Universe is infinite, we are surrounded by a boundary. Any galaxy so far away that light from this galaxy requires more than 13.7 billion years to reach us is beyond the *horizon*. A galaxy beyond the horizon is outside the causal web; there has not been time since the Big Bang for anyone in this galaxy to have affected us in any way or for anyone in this distant galaxy to have been affected by us.

Four hundred thousand years after the Big Bang the horizon was much closer. The uniformity of the cosmic background radiation is remarkable, because the radiation we see on opposite sides of the sky is from places that were not within each other's horizon at this time. A homely analogy is making a cake. On reflection, I have actually never made a cake, but I think I saw one made once on television. If I remember correctly, the cake mix was initially quite lumpy, and it was only by constant stirring that the cook managed to get out most of the lumps. There is no particular reason, in the standard Big Bang theory, to expect that the Universe at the moment of the Big Bang – the cake mix – should have been smooth. The Universe might actually have been very lumpy at that time. But the uniformity of the cosmic background radiation shows that it was clearly remarkably well-mixed only 400,000 years later. This is a paradox, because there had not been time for all the different places we observe through the cosmic background radiation to have been affected by each other. There was not time to mix the cake.

The cake-mix problem (often more boringly called the *horizon problem*) had nagged at people's minds since the discovery of the cosmic background radiation. In 1981 an American physicist, Alan Guth, suggested a possible solution. Guth's theory of inflation was more a rough sketch than a fully worked out scientific theory, but it was a sketch that neatly solved the problem.

Before the advent of quantum theory, physicists explained the forces of nature as being produced by smooth *fields* that fill space. The "classical" view of the electrical force exerted by an electron, for example, was that the electron is surrounded by an electric field, and it is this field that produces the force. As I described earlier, quantum theory implies we should instead visualize fields, albeit through a glass darkly, as being composed of virtual particles, ghostly particles which flicker in and out of existence. It is fairly obvious that fields often contain energy (a dropped glass is enough to show that a gravitational field contains energy), but a surprising prediction of the quantum theory is that there may be a minimum energy that a field can contain. In the classical view, the minimum energy is zero, but the quantum view is that the ghostly particles that constitute a field may set a lower limit to its energy – the *vacuum energy*.

In their attempts to unify the forces of nature, the authors of the grand unified theories had suggested that there might have been additional fields in the Universe shortly after the Big Bang. Guth realized that the vacuum energy contained in one of these early fields might be the solution to the cake mix problem. He proposed that about 10^{-34} seconds after the Big Bang, the vacuum energy in one of these fields caused the Universe to go through a period of rapid expansion, inflating it like a balloon. One current estimate is that the Universe may have expanded by a factor of 10^{27}, a one followed by 27 zeros. As the Universe expanded, the energy in the field fell, and eventually the period of *inflation* came to an end*.

* This may sound like the dark energy I described in Chapter 8, but the amount of dark energy is minuscule compared to the energy in one of these early fields. The astute reader may wonder how the energy fell if the vacuum energy is the minimum energy that a field can contain. This is one of the many parts of the inflation theory that is still controversial and it is why inflation is more a sketch of a theory than a fully worked out scientific theory.

This colossal expansion factor provided a natural solution to the cake mix problem. The region of the Universe we can observe, the region enclosed by the horizon, is huge; the current distance to the horizon is 45 billion light years*. Nevertheless, if Guth's idea is true, the expansion that occurred during the period of inflation was enough to produce this immense volume of space from something smaller than the size of an atom. Guth realized that 400,000 years after the Big Bang, two places, which according to the standard Big Bang theory had never been in causal contact, might actually have been in contact before the great expansion. Consider the cake mix before it has been stirred by the cook. Although it contains big lumps, the cake mix within the individual lumps has probably already evened itself out without any work by the cook. The basic idea of inflation is that the volume of space we can observe did not arise from a whole pan of lumpy cake mix, but from a minuscule speck within the pan. If true, the uniformity of this volume is not so surprising.

Guth's idea also provided a way to make the sound waves we observe in the Universe 400,000 years later. Because of the graininess in the field produced by the virtual particles, there must have been minuscule variations in the energy of the field from place to place. Inflation would have greatly amplified these variations because it would have ended at places where the energy was higher than average at a slightly later time than at places where the energy was below average. Theorists have shown that these *quantum fluctuations* would naturally have produced the frequencies of the sound waves that we observe.

As an observer, I must admit, I never believed any of this. For almost two decades after the publication of Guth's paper, cosmologists were divided into two camps. The theorists generally liked

---

* Distance is a treacherous concept in an expanding Universe. The light from an object on the horizon has been travelling for 13.7 billion years. We can calculate that the current distance of this object from the Earth must be about 45 billion light years, but of course we do not know anything about this object *today* – we only know what it looked like 13.7 billion years ago.

Guth's idea. Observers tended to think of inflation as a crazy theoretical idea, for which there was no evidence, and which it would never be possible to test. The thing that changed my mind was BOOMERANG.

At this point, we must revisit Fred, the 2D creature living on the surface of a sphere. Remember that it is possible for Fred to find out that he is living in a curved universe by carrying out a geometrical test, as long as he carries out the test over a large enough area for the curvature to be evident. If he adds up the angles in a triangle, but the triangle is much smaller than his universe, he will just discover the exciting fact that the sum of the angles is 180 degrees. Let us suppose that Fred's universe has passed through a period of inflation during which it has expanded by a factor of 10^{27}. It is likely that the part of his universe that he can observe, the part that is within his observational horizon, has swollen up from something less than the size of an atom. In which case, his entire observable universe will be less than a speck in Figure 9.2. If this is true, he will never be able to carry out a geometric test over a large enough part of his universe to tell it is curved. As it is with Fred, so it is with us. If inflation actually occurred, our entire observable Universe must only be a speck

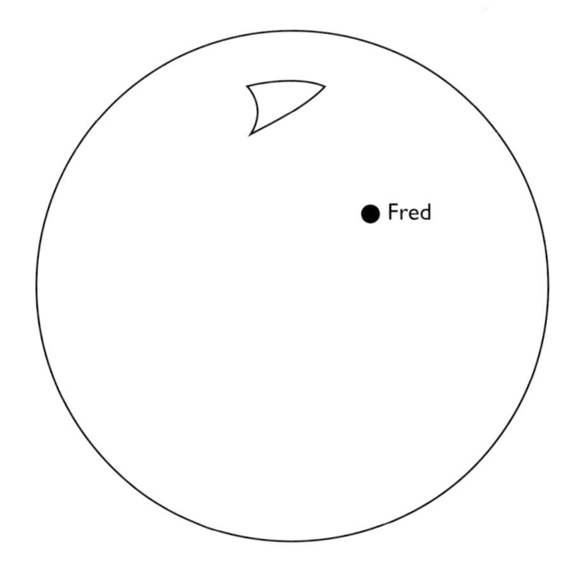

FIGURE 9.2 Fred revisited – a 2D observer in a 2D universe.

within the greater Universe. The theory therefore makes the prediction that any geometric test, even one like BOOMERANG that spans the observable Universe, will just show that space is flat. This, of course, is just what was discovered by the BOOMERANG team. The BOOMERANG result and the fact that the theory does provide a way of generating sound waves with the correct frequencies have convinced even this sceptical observer that this remarkable event probably happened.

Inflation is an awesome theory. In Chapter 4 I tried to give an idea of the sheer emptiness of the Universe. Blaise Pascal, the seventeenth century mathematician and theologian, had this reaction to the barely fathomable distances between the stars: "The eternal silence of these infinite spaces frightens me." I never felt the same as Pascal until very recently, because even if the distances to the stars, let alone the distances to the galaxies, are incomprehensible in human terms, yet we can observe these objects with our telescopes. We can sit here on the third planet of an average star and observe objects over ten billion light years away – and for me this always brought these incomprehensible distances back on to the human stage. Inflation destroyed this cosy feeling. We may be able to span the region within the horizon with our telescopes, but if the theory is correct, this region is merely a mote within the greater Universe. The Japanese poet Issa compared our human world to the world of micro-organisms visible within a drop of dew, but a metaphor that gets closer to the truth of the matter is if the dewdrop contains our entire observable Universe ninety billion years in diameter.

The surface of the dewdrop is the horizon. By definition, we can not observe anything beyond the horizon, but is there any way at all of finding out what is happening in the wider world outside our dewdrop? The answer is: just possibly.

To see why there just might be a way to do this, let us return to the set of big questions in Figure 9.1. My undergraduate course on the Big Bang theory – and this book so far – contains the answer to only one of these questions (question 4, as it happens). Nevertheless, over the last three decades, scientists have tried to broaden the scope of the standard scientific method to address more of these big questions. They still cannot answer the fundamental questions about life that keep us awake at night, but they have at least got plausible (although not necessarily correct) answers to two of the nonexistential questions in the list: questions 9 and 14.

They have found these answers by applying the scientific method, which is really just refined common sense, to the one fact among millions that is the most interesting thing about the Universe: the fact it contains us. As I know I exist (I am not so sure about you – see question 5), I can deduce the Universe must have certain properties, because otherwise I would not be here. This idea was first stated by the physicist Brandon Carter in 1974 as the Anthropic Principle: *the Universe must have certain properties for human beings to be here to observe it**. This sentence sounds too obvious to even call it common sense and, as I will explain below, it is actually a can of philosophical worms. However, first let me show the Anthropic Principle in action, which is the easiest way to explain it.

Carbon is the magical element. The chemical properties of the carbon atom, in particular its four chemical bonds, make possible the complex molecules on which all life is based. Like most other elements, carbon is made in the nuclear furnace in the centers of stars (Chapter 4). However, in the early 1950s when the physicist Fred Hoyle began to study the details of the nuclear reactions that form carbon, he realized that the fact it exists at all is rather surprising. A carbon nucleus is formed when three helium nuclei fuse together. However, when Hoyle made careful calculations, he discovered that the chance of such a three-nucleus encounter is extremely small. He realized that for there to be such a large amount of carbon in the Universe, the probability of three nuclei fusing when they do approach each other must be extraordinarily high. He realized that for this probability to be so high the carbon nucleus must have a special internal structure that enhances the probability of one nucleus fusing with another. All of the supporters of anthropic reasoning always quote this example

* I have paraphrased Carter's original definition, but the meaning is the same. Carter coined the term *Anthropic Principle* from the word *anthropocentric*. The latter word, of course, means judging things from a narrow human perspective. This is usually thought to be a rather bad thing. Carter decided that for cosmological research it is sometimes a good thing.

because of the success of Hoyle's deduction. A few years later, when physicists finally measured its structure, they discovered that the carbon nucleus has exactly the features predicted by Hoyle. This is an example of anthropic reasoning because of the following sequence of arguments: we exist; we only exist because there is a large amount of carbon in the Universe; for there to be so much carbon, the carbon nucleus must have a particular internal structure.

A second example is the strength of gravity. The gravitational attraction between a proton and an electron is a factor of 10^{39} times less than the electrical attraction between the two particles. Gravity only wins out on the scale of planets because a planet contains equal numbers of positively and negatively charged particles; the electrical repulsion between particles with charges of the same sign balances the attraction between particles with charges of the opposite sign. It is easy to show that in an alternative universe in which the gravitational force was rather stronger relative to the electrical force, we might not exist. In an elegant book on the Anthropic Principle, *Just Six Numbers*, Martin Rees has argued that in this alternative universe stars would be formed at a much faster rate, but they would also shine more brightly, rapidly exhausting their fuel and not leaving enough time for life (and us) to evolve. The ratio of the gravitational and electrical forces is one of the eponymous six numbers: six basic properties of the Universe on which our existence appears to depend. If any of these numbers were slightly different, so Rees argues, the human race would not exist – for a variety of reasons, including a lack of galaxies, planets, chemical ingredients and time.

A second of these numbers is the dimensionality of space. The origin of the argument for this number predates Rees, Carter or any of the modern supporters of anthropic reasoning. In the eighteenth century, the clergyman William Paley in his book *Natural Theology* showed that the orbit of a planet is only stable if the gravitational force between the star and the planet decreases as the square of the distance between the two – the form of Newton's famous law of gravity. Paley argued that if the form of this law were any different, the planet would either spiral into the star or escape from it completely. Since Paley's time, many scientists have argued that the form of Newton's law is the consequence of the dimensionality of space, and that a universe with more or less than three dimensions would have a law of gravity with a different form[21]. The fact that we live in a universe with three dimensions is therefore not surprising. If space had a different number of dimensions, planetary orbits would not be stable – and we would not be here to think about it.

Thus, if the supporters of anthropic reasoning are correct, we now have answers to two more of the big questions in Figure 9.1. Space has three dimensions and the constants of nature have the values that they do, because otherwise we would not be here.

Do these answers satisfy you? They should not. Hidden in these apparent answers to two big questions is an even bigger one. If space had a different number of dimensions, if gravity were rather stronger than it is, if the constants of nature had different values – I would not be here typing these words and you would not be here reading them. But what is the significance of the fact that the Universe appears to be well-fitted for our existence? There are at least four possible answers[22]:

1) ***The Design Answer***: This was the answer proposed by Paley in *Natural Theology*. God is a cosmic watch-maker who designed the Universe as a home for humans – and so it is not surprising that it is a nice place to live.

2) ***The Anthropocentric Answer***: Anthropic reasoning is too anthropocentric. We are being too limited in our imagination. The human race might not have evolved if the gravitational force had been rather stronger, but some kind of intelligent life would probably have evolved, even if it is beyond our abilities to imagine what it would have been like.

3) ***The Multiverse Answer***: There are many different universes, each with a different set of values for the constants of nature, and possibly with a different number of dimensions. Life evolves in some universes, but not in others. We obviously find ourselves living in one which is well-suited to us.

4) ***The Grumpy Observer Answer***: The fact that the Universe is well-suited for our existence has no deep significance at all. Even if we had never looked at the sky through a telescope, we could have been certain that this is just what we would find once we started to study the Universe. We should therefore not try to make any deductions at all from something that is so obvious. In philosophical terms, the Anthropic Principle is a tautology.

If we just consider hard scientific data, the kind we collect with telescopes and particle accelerators, none of these answers, at present, is any better than any other. You are free to choose whichever answer suits your taste. You may be a religious fundamentalist and prefer the first or you may be a science fiction fan and like the second because it implies our own kind of life is not unique. I may disagree with you, but I have no hard evidence to back up my preference. In the absence of data, the answer you prefer will inevitably be based on your cultural, religious and philosophical preconceptions. We are now clearly in the realm of philosophy and religion rather than science. Even looking to the future, it seems to me that there is only one of the four answers that scientists may ever be able to test.

The multiverse is a by-product of the attempts to invent a theory that unifies all four forces of nature. Theorists are well-known for the prodigality with which they invent and discard theories (sometimes even before morning coffee), but one type of unification theory that has endured through the ins-and-outs of theoretical fashion are string theories. In these theories, reality has far more than the four dimensions (time and the three space dimensions) that we observe. Different string theorists propose different numbers of dimensions, but for reasons too complex to explain the minimum number is ten. The extra dimensions are unobservable in the same way that the third dimension cannot be observed by Fred, the observer in the 2D universe. In the string world, in which ultimate reality has 10, 11 or even 25 dimensions, it no longer seems ridiculous to imagine a universe like ours but with five or six dimensions.

Our entire observable Universe, according to the theory of inflation, was produced from a tiny seed. If inflation happened once, it is possible that it happened many times, with different universes inflating from different minuscule seeds within the greater *multiverse*. If string theories or some of the other unification theories are correct, it is possible that these different universes ended up with different physical constants and different numbers of dimensions. The multiverse answer to why we live in such a benign universe is that while some universes will be barren of life, others will be suitable for life – and ours obviously will be supremely well-suited to our kind of life.

The multiverse idea is an interesting one because it is at least an attempt to reason about the wider world outside our dewdrop. For an observer, however, it is a rather frustrating concept because it is hard to see how we will ever be able to test it.

Nevertheless, I do not want to finish this book on a depressing note. The number one rule for observers, ingrained in the soul after countless observing runs lost to the weather or broken instruments, is: never give up and look to the future. So let me finish this book by describing an experiment aimed at testing an important part of the multiverse idea: *inflation*.

The truth of one does not imply the truth of the other. The Universe may have undergone a period of inflation without there necessarily being other universes with different numbers of dimensions and different values for the physical constants. As I described earlier, the flatness of space and the particular music heard shortly after the Big Bang are persuasive evidence that inflation did occur. However, is there any way of finding more conclusive evidence? Remember that inflation happened (if it did happen) at a time only 10^{-34} seconds after the Big Bang.

One way is to look for gravitational waves. Electromagnetic waves are waves which travel through the electromagnetic field and include all the varieties of radiation I have described in this book. I have so far not mentioned gravitational waves, which are waves that travel through the gravitational field, for the very good reason that nobody has yet detected them. Nevertheless, the first gravitational-wave telescopes are now in operation, even if they have not yet detected anything, and we are on the verge of the era of "gravitational-wave astronomy." A gravitational-wave telescope, however, looks very unlike a conventional telescope. For a start, it is underground (to shield it against vibrations) and, instead of a mirror, it consists of a system of masses, which are linked by laser beams and which vibrate when a gravitational wave passes through the telescope.

Electromagnetic waves are produced by the motion of electric charges; gravitational waves are produced by the motion of masses. The motion of the Earth around the Sun is producing gravitational waves, although they are far too weak to detect. The fluctuations in the quantum field 10^{-34} seconds after the Big Bang, the ultimate cause, according to the theory of inflation, of the lumpy Universe we see today, also produced gravitational waves. The limit to the powerful technique of observing history, which has been one of the major themes of this book, is that before 400,000 years after the Big Bang electromagnetic photons were scattered by electrons. Gravitational waves however are *not* scattered by electrons. Therefore, the gravitational waves from the period of inflation can reach us, and so, in principle, we can use them to observe the history of the Universe to within a split second after the Big Bang. This technique holds huge promise for the future, but unfortunately current gravitational-wave telescopes do not have the sensitivity to detect gravitational waves from this period (or even, so far, from anything).

Until gravitational-wave telescopes come of age, it may be possible to detect these waves indirectly. The gravitational waves generated during the period of inflation have been travelling through the Universe ever since, and because they are just waves in the gravitational field they will affect objects in the same way that gravity does. As they travelled through the Universe 400,000 years after the Big Bang, they will have perturbed the gas that the cosmic background radiation emitted. Theorists have predicted that the gravitational waves will have produced a subtle effect on by this radiation. Electromagnetic radiation usually consists of an equal mixture of two polarized waves. One way to understand this is to think of an electromagnetic wave as being like a wave on a string: one polarization consists of oscillations of the string in the vertical direction and the other of oscillations in the horizontal direction. According to the theorists, the effect of the gravitational waves passing through the Universe 400,000 years after the Big Bang will have been to produce a tiny difference in the two polarizations of the cosmic background radiation. The theorists predict that the size of this difference will vary over the sky in a distinctive pattern. This pattern will be extremely difficult to detect. The observations with COBE, BOOMERANG and WMAP that I described in the last chapter showed that the cosmic background radiation varies from place to place across the sky by about one part in 100,000. The variation in the cosmic background radiation produced by the gravitational waves is many times smaller than this.

Nevertheless, ever optimistic, scientists around the world are planning new telescopes to look for this effect. The group that I know best is planning to construct a telescope in Antarctica. The telescope will observe in the submillimeter waveband, in which this effect is expected to be strongest. Because submillimeter observations are limited by the amount of water vapor in the atmosphere (Chapter 5), the group plans to construct this telescope in one of the driest (and most inhospitable) places on Earth: Dome C, one of the peaks on the Antarctic Plateau, where the observing conditions are expected to be even better than at the South Pole. The telescope, which will be called CLOVER (an acronym too complicated to explain), should be observing by 2008. If CLOVER detects the predicted pattern, this will be firm evidence that inflation actually happened. Of course, CLOVER may not detect anything, and even if CLOVER shows that inflation did happen, this does not necessarily mean that the multiverse exists.

But it is a start.

Notes

1. Pickett, M.K. and Lim, A.J. (2004), *Astronomy and Geophysics*, vol. 45.
2. Laskar, J. and Robutel, P. (1993), *Nature*, **361**, 608; Laskar, J., Joutel, F. and Robutel, P. (1993), *Nature*, **361**, 615.
3. Correia, A.C.M. and Laskar, J. (2001), *Nature*, **411**, 767.
4. Peale, S.J. (1999), *Annual Reviews of Astronomy and Astrophysics*, **37**, 533.
5. Carnup, R. (2004), *Annual Reviews of Astronomy and Astrophysics*, **42**, 441.
6. Calculation by the author.
7. Giotto mission website (European Space Agency).
8. Tombaugh, C.W. (1991), *Sky and Telescope*, **81**, 360; Tombaugh, C.W. (1996), *Astronomical Society of the Pacific Conference Proceedings*, **107**, 157.
9. Luu, J. and Jewitt, D., *Scientific American*, May, 1996.
10. Drake's reminiscences of Project Ozma are given in *Cosmic Search Magazine* (1979, Vol. 1, No. 1), a short-lived magazine whose remains can now be found on the web (www.bigear.org/vol1no1/ozma.html).
11. The idea of using the frequency of the hydrogen line for contacting extraterrestrials is usually attributed to Cocconi and Morrison, who suggested it in a paper in the scientific journal *Nature.* However, it is clear from Drake's reminiscences that he thought of this independently.
12. Gomes, R.S., Morbidelli, A. and Levison, H.F. (2004), *Icarus*, **170**, 492.
13. Morgan, H., Dunne, L., Eales, S., Ivison, R. and Edmunds, M. (2003), *Astrophysical Journal*, **597**, L33.
14. Wynn-Williams, C.G. (1982), *Annual Reviews of Astronomy and Astrophysics*, **20**, 587.
15. Calculation by the author.
16. Hu, W. and White, M. *Scientific American*, February, 2004.

17. Calculation by the author based on the results of Fixsen *et al.* (1998), *Astrophysical Journal*, **508**, 123, and Dwek *et al.* (1998), *Astrophysical Journal*, **508**, 106.
18. Gamow, G. (1946), *Physical Review* **70**, 572; Alpher, R.A., Bethe, H. and Gamow, G. (1948), *Physical Review* **73**, 803.
19. Calculation by the author from the range of cosmological parameters in vogue in 1995.
20. Spergel, D.N. *et al.* (2003), *Astrophysical Journal Supplement*, **148**, 175.
21. Section 4.8 in *The Anthropic Cosmological Principle* by John Barrow and Frank Tipler.
22. I have adapted these from those given by Anthony Aguirre in *On making predictions in a multiverse: conundrums, dangers and coincidences* (astro-ph 0506519).

Bibliography

I did most of my research for this book by reading research papers and reviews. Outside my own research area, I found review articles in the journal *Science* and in *Annual Reviews of Astronomy and Astrophysics* particularly useful. I have only cited individual articles when I think my colleagues might wonder where I got the information or find a point contentious. These references are given in the notes. Apart from articles, I found the following books particularly useful.

Alvarez, Walter. *T. Rex and the Crater of Doom*. Penguin Books, 1999.

Barrow, John D. and Tipler, Frank J. *The Anthropic Cosmological Principle*. Oxford University Press, 1996.

Brogan, Hugh. *The Pelican History of the United States of America*. Pelican Books, 1987.

Carey, John (editor). *The Faber Book of Science*. Faber and Faber, 1995.

Christianson, Gale E. *Edwin Hubble, Mariner of the Nebulae*. University of Chicago Press, 1996.

Christianson, John Robert. *On Tycho's Island: Tycho Brahe, Science and Culture in the Sixteenth Century*. Cambridge University Press, 2003.

Galilei, Galileo. *The Sidereal Messenger*. 1609.

Hubble, Edwin. *The Realm of the Nebulae*. Yale University Press, 1936.

Johnson, George. *Miss Leavitt's Stars – the Untold Story of the Woman Who Discovered How to Measure the Universe*. W.H. Norton, 2005.

Koestler, Arthur. *Sleepwalkers: A History of Man's Changing Vision of the Universe*. Hutchinson, 1959.

Longair, Malcolm. *The Cosmic Century: A History of Twentieth Century Astrophysics and Cosmology*. Cambridge University Press, in press.

Morton, Oliver. *Mapping Mars*. Fourth Estate, 2002.

Munitz, Milton K. (editor). *Theories of the Universe*. The Free Press, 1957.

North, John. *The Fontana History of Astronomy and Cosmology*. Fontana Press, 1994.

Peacock, John. *Cosmological Physics*, Cambridge University Press, 1999.

Peebles, P.J.E. *Principles of Physical Cosmology*. Princeton University Press, 1993.

Rees, Martin. *Just Six Numbers: The Deep Forces that Shape the Universe.* Phoenix, 2000.

Shapley, Harlow. *Through Rugged Ways to the Stars*. Charles Scribner, 1969.

Sobel, Dava. *Galileo's Daughter*. Fourth Estate, 1999.

Stephenson, F.R. and Green, D.A. *Historical Supernovae and their Remnants*. Oxford University Press, 2002.

Tico Brahae. *His astronomicall coniectur of new and much admired [star] which appeared in the year 1572*. Early English Books Online.

Winchester, Simon. *The Map that Changed the World*. Viking, 2001.

Whitney, Charles A. *The Discovery of Our Galaxy*. Alfred A. Knopf, 1971.

Wright, Helen. *Explorer of the Universe: A Biography of George Ellery Hale*. E. P. Dutton & Co., 1966.

Index

Printed in China